D1067819

BEST-KEPT SECRETS OF THE

GREEK ISLANDS

Publisher and Creative Director: Nick Wells
Project Editor and Picture Research: Cat Emslie
Art Director and Layout Design: Mike Spender
Digital Design and Production: Chris Herbert
Copy Editor: Anna Groves
Proofreader: Dawn Laker
Indexer: Helen Snaith

Special thanks to: Victoria Lyle and Megan Mizanty

18

7 9 10 8 6

This edition first published 2009 by
FLAME TREE PUBLISHING
6 Melbray Mews, Fulham,
London SW6 3NS

www.flametreepublishing.com

ISBN 978-1-84786-648-6

The photographs are courtesy of the following picture libraries and © the following photographers:
Alamy and Roger Cracknell 08/Greece: 143, 180–81; Chris Deeney: 38–39; Greg Balfour Evans: 18; Robert Harding Picture Library Ltd: 182; Terry Harris Just Greece Photo Library: 55, 166–67, 168, 169; IDP Greek Collection: 189; IML Image Group Ltd: 44–45, 48–49, 66, 67, 88–89, 96; Andre Jenny: 80–81; Rainer Kiedrowski/Bildarchiv Monheim GmbH: 133; Jason Knott: 52, 53; Roberto Meazza/IML Image Group: 116; Patrick Medd: 28–29; Peter Molz: 155; Pictures Colour Library: 14; Nicholas Pitt: 25, 183; Ellen Rooney/Robert Harding Picture Library Ltd: 17; **Fotolia** and Ageless Adventurer: 161; Richard Coltart: 126; elimare: 59; hattiney: 92; Axel Kramer: 140; Oneworld-images: 63; Pete Oraica: 153; Pavlos: 79; Sebastian Schulz: 74, 75; skkh: 119; Vangelis Thomaidis: 141; Tony Z: 46; **iStockphoto** and Matjaz Boncina: 117; Emmanouil Filippou: 132; Constantinos Gerakis: 142; David Newton: 185; Saso Novoselic: 159, 160; Gianluca Padovani: 154; Pidjoe: 136; Margaret Rendle: 127; Marco Testa: 77; **Hugh Palmer**: 30, 32, 62, 71, 78, 82, 87, 93, 120; **Photolibrary** and Mark Banks: 22–23; Walter Bibikow: 164; Frank Chmura: 58; Yiorgos Depollas: 103, 156–57, 158, 165; Guiziou Franck: 65; Ken Gillham: 147, 152; Fraser Hall: 20; Robert Harding: 24, 178–79; Jeremy Lightfoot: 162–63; Doug Pearson: 16; Juergen Richter: 109; R H Productions: 21, 176–77; Ellen Rooney: 64; **Shutterstock** and airphoto.gr: 144–45; alex&alexL: 31; Georgios Alexandris: 26–27, 40–41, 54, 60, 68, 134–35, 138–39; Ariy: 47; baldovina: 146, 148–49; Josef Bosak: 35, 36–37; Aron Brand: 124; casinozack: 188; Paul Cowan: 111, 115; Gary Dyson: 33; easyshoot: 84–85; efilippou: 130; Goncalo Veloso de Figueiredo: 118; filarx3: 173; Babusi Octavian Florentin: 184; Karel Gallas: 131; George Green: 15; Andrey Grinyov: 61; Mirek Hejnicki: 106–07; Marcel Jancovic: 137; Kert: 102; Andrey Kudinov: 112; R.J. Lerich: 73; Mauritania: 113; Holger Mette: 114; Netfalls: 19; newphotoservice: 69, 70, 72, 110; Morozova Oksana: 13; P. Phillips: 34, 51, 121; Pixage Photography: 12; Asta Plechaviciute: 122–23, 129; Robert Ranson: 91; Jane Rix: 50; Dimitrios Rizopoulos: 186–87; Jan Schuler: 83; Juliya W. Shumskaya : 108; sima: 172; Svetlana Tikhonova: 97; Petros Tsonis: 76; ultimathule: 86, 128; vlas2000: 170–71; Dmitry Yatsenko: 125; Alex Yeung: 94–95, 98–99, 100–01; Maria Yfanti: 90.

Printed in China

BEST-KEPT SECRETS OF THE

GREEK ISLANDS

Diana Farr Louis

FLAME TREE PUBLISHING

CONTENTS

CONTENTS

INTRODUCTION

It is no secret that the islands of Greece are among the most spectacularly beautiful places on the planet. Even if you have never been to the Greek islands, you have seen their idyllic images on posters and brochures, and you have watched them unfold as romantic settings in dozens of films. The notion of these islands seems to embody all our longings: the desire to get away from it all, to be seduced by natural splendour, to shed our inhibitions – perhaps our clothes – on a glistening beach, to spend the night under the stars, to watch the sun rise out of the Aegean and to notice that the dawn does indeed have 'rosy fingers', just as Homer said.

People come to Greece for reasons as numerous as there are islands; to warm their bones and worship the sun after months of Northern darkness, to party after a year of studies or office chores, to pay homage to ancient civilizations or to discover a bit about a modern one. Some like to island hop and some return year after year to the same village, while others never leave their armchairs but wander freely in their imaginations.

The 140 photographs in this album are aimed at giving you an impression of all the large and many of the smaller of Greece's 227 inhabited islands. The captions accompanying them attempt to distil their essence, an almost impossible task in so few words. While we all know a picture is worth a thousand words, the chosen images and the text can but hint at the islands' secrets.

To orient the reader, we present the islands in geographical order as if they were revolving anti-clockwise around the Greek mainland, starting with the Ionian archipelago off the west coast, hopping east to the Argo-saronic islands, then proceeding south down the middle of the Aegean to the Cyclades, on to Crete and round to the eastern Aegean and the Dodecanese, then up to the North-east Aegean group as far as the Dardanelles, before winding up with the Sporades and Evia, off the coast of Thessaly, Viotia and Attica.

You will notice that some features are constant: brightly coloured kaikis (fishing boats) bobbing on shimmering water, little houses clustered around a harbour, a profusion of flowers, churches and castles, olive groves, cypress trees, hillside villages and, above all, that celebrated Greek clarity of light. All appear throughout the islands, regardless of location or size. But beyond these similarities, each island, no matter how insignificant, will have its own character, its own individual personality, something that may not show up in a photograph but that you have to experience first hand and with all your senses.

We can only reveal some of the secrets visible to the naked eye or the camera lens. We cannot bring you the smells: of a jasmine-perfumed terrace on Spetses, of thyme as you walk on a hillside on Folegandros, of anise-scented ouzo enjoyed in a waterfront café, of oregano freshly sprinkled on a slab of feta, of Corfu's intoxicating wild strawberries, of sulphur fumes emitted by hot springs on Milos and Nisyros (just to mention two) or of chops sizzling on charcoal outside a rustic taverna.

We cannot summon up the tastes: of a grilled octopus tentacle eaten on the tiny, crowded waterfront at Naoussa on Paros, of a Santorini tomato grown without water, of sea spray on your lips as you cross from Mykonos to Delos in a *meltemi*, of fruity extra virgin olive oil poured on a brick-like rusk in Crete, of a sweet, wrinkly Thasos olive or a sharp, wine-soaked Kos cheese, or even a chewy *loukoumi* from Syros, flavoured with rose water. Aigina's pistachios, Lesvos's impeccably cooked catch of the day, Sifnos's chickpea casseroles, Chios's mastic-flavoured liqueur are all missing from this album.

And what about the sounds? Of goat bells tinkling on a mountainside, or church bells clanging (way too early) on a Sunday morning; the chug-chug of kaiki engines coming back from fishing, gulls screeching above them; a too loud television

showing a soccer game and the roars at the café when the right team scores; the hornet's buzz of motorbikes (the bane of some islands); and of course music, music everywhere, from the plunkety-plunk of a bouzouki combo to Julio Iglesias, Madonna and Greece's own heart-throb, Sakis Rouvas, Eurovision hero.

One of the most important sounds is the Greek language itself. It is rarely whispered. Conversations are conducted at decibels higher than anywhere in Europe except for Naples (which was founded by Greeks). Perhaps that's a consequence of shouting across mountains or storms at sea. Greeks love noise and seem to be happiest when making themselves heard above the din in a taverna or nightclub. People do not appear in our photos, but they cannot be eliminated from any introduction to Greece.

Islanders are apt to be exuberant, intelligent, warm, curious and helpful, but much easier to approach in winter than in summer, now that so many of them work in the tourist trades and get exhausted by August.

Tourism is having an impact not only on the people but also on the appearance of the islands to a degree that has occurred only rarely in the past, and never as rapidly. Ironically, many of the features that make these places so attractive – castles and fortifications, mosques and minarets, palaces and quaint mountain villages – were actually the products of conquest and exploitation by Western Europeans, Turks and pirates over several centuries. It remains to be seen whether time will put such a romantic patina on today's reinforced concrete hotels and tourist ghettoes.

In 1204, Catholic Crusaders led by a vindictive Doge mounted an attack on Constantinople. Instead of voyaging to oust Muslims from the Holy Land, they turned on their fellow Christians, replaced the Orthodox Byzantine emperor with a Frankish king and occupied virtually all the islands. Frankish noblemen erected castles at strategic points, usually on the site of the ancient acropolis, and treated the islands as their fiefs and the islanders as serfs. Gradually Venice, the supreme naval/trading power of the era, assumed control of all but the few islands that her arch rival Genoa had captured. As a result, there is not a single major island without a medieval castle with walls to offer some degree of protection to its inhabitants.

In 1453, the Ottoman Turks conquered Constantinople and began a tug of war with Venice for trading routes and possessions. By 1715, they had replaced the Venetians as rulers on every island except those in the Ionian. Meanwhile, in 1538, Barbarossa, a particularly venomous pirate in the service of the Turks, swept through the Aegean, virtually exterminating whole populations of islanders from Spetses to Skiathos. Years later, when the few timid survivors eventually returned, they built those emblematic white villages – usually called Hora or 'main place' – as far inland and as inaccessible as possible in the Cyclades, Sporades and Dodecanese. They crowded their houses along warren-like alleyways to confuse would-be intruders. Outside the walls, they carved the stony hillsides into strips of cultivable soil and 'planted' chapels by the hundreds when prayers were answered or a life spared. When new urban buildings went up, after Independence in the nineteenth and early twentieth centuries, they kept to a harmonious neoclassical style.

Since the 1960s, the landscapes of some Greek islands have been changing once more. Hotels, immense and petite, hideous and elegant, rise on beachfronts thought worthless by the farmers who previously owned them. Fishing villages have burst from their harbours, tacky complexes and bungalows insult formerly pristine coasts. Much modern construction does not respect local proportions and traditions. Prefab chapels adjoined to three-storey villas with cement courtyards are a secret we'd rather keep to ourselves, along with rubbish dumps, burnt forests and plastic-littered sands. We cannot blame foreigners and tourism for all of Greece's lapses. This time, it is not war but peace and prosperity that are altering the scenery.

That said, these islands continue to fascinate and beguile. I never cease to be amazed at the persistence of 'Greekness'; how islands as far apart as Corfu and Kastellorizo spoke the same language, had the same gods and myths, the same sense of 'homeland' despite their differences in heritage and history. I love the overlap of past influences: obsidian chips, Classical pottery shards and a Roman coin protruding from layers of eroded wall on an Andros or Kythera beach; a fortress with column drums in its bastions; a nineteenth-century chapel decorated with early Christian and ancient reliefs. I love too the sight of red anemones against old marble, diving into silken water, chatting in a taverna for hours without being asked to move on, the simple, true tastes of good Greek cooking, islanders' kindness and *kefi* ('joie de vivre') and many more things too numerous to mention. Most of all, I appreciate being surprised. This country is never boring. If it were all perfection, I think I would have tired of that island dream long ago.

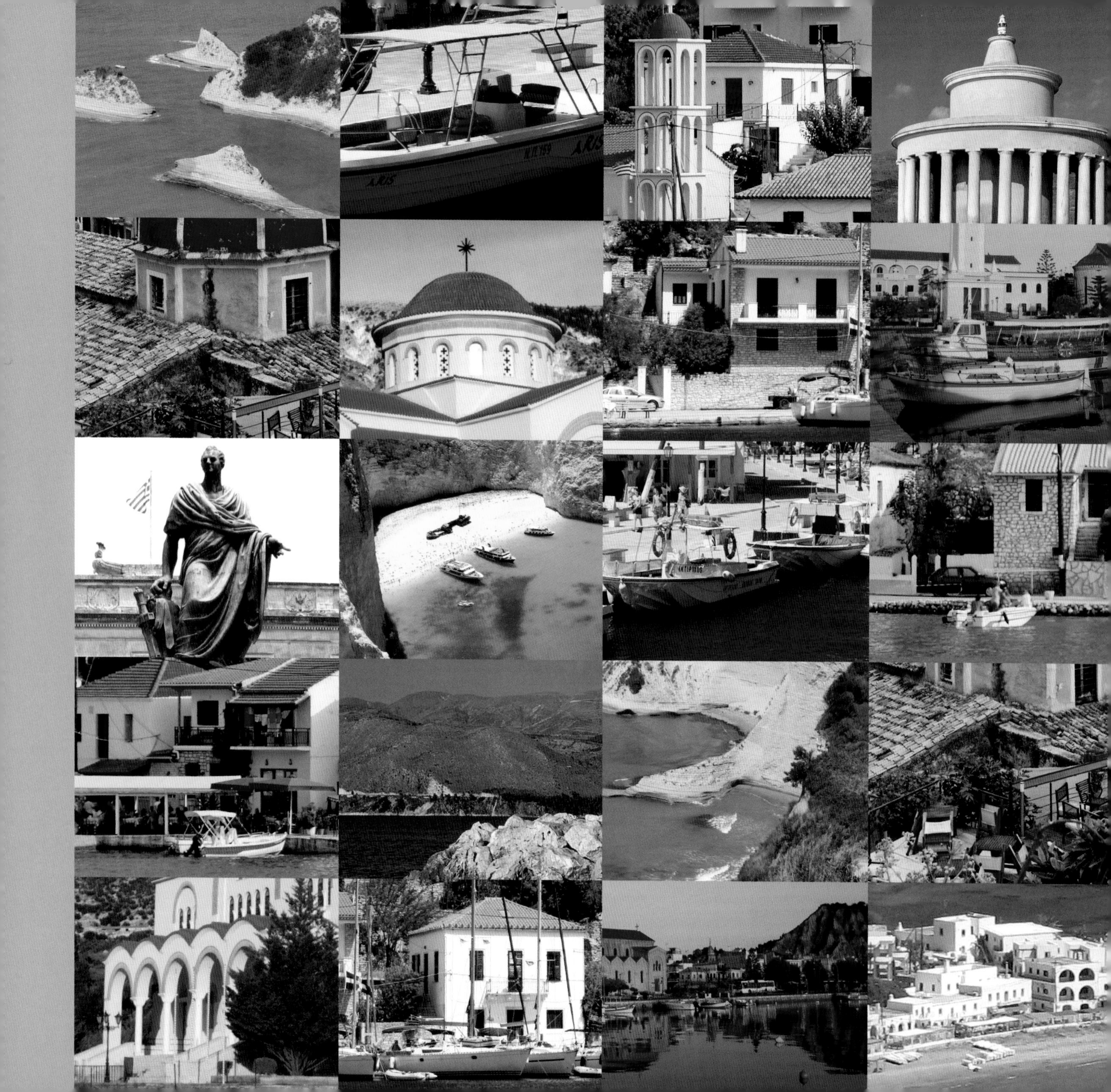
ARIS

THE IONIAN ISLANDS

Greece's 'emerald isles', so much lusher and greener than their Aegean counterparts, are set in the Ionian Sea between Italy and the Greek mainland. The Greeks call them 'the Seven Islands', but this refers only to the seven largest – Corfu, Paxos, Lefkada, Ithaka, Kefallonia, Zakynthos and Kythera. In reality, there are dozens more, inhabited and not, while Kythera, lying between Crete and the eastern prong of the Peloponnese, has no geographic connection whatsoever to the others.

Instead, the number reflects the group's history as a collective trophy, won successively by the Venetians, French and British, who held the islands for almost 600 years, from the thirteenth to the nineteenth century. This intimate and prolonged contact with the West, combined with the almost total absence of Ottoman influence, left indelible traces on their architecture, language, cuisine, music and even landscape (think Corfu's olive groves and Zakynthos's currant vines).

Today these beautiful, gentle islands play host to a new wave of Western 'invaders' – Italians, French and especially British tourists who have also left their mark on their welcoming shores. But the proliferation of 'rooms for rent', baked beans and English pubs has not tarnished the hinterland and the Ionians still conceal a myriad of unspoiled charms.

CHURCH DOMES IN CORFU TOWN

Corfu (Kerkyra)

Red domes abound in the old town of Corfu, whose cobbled streets, tiled roofs, faded pastel tenements and fluttering laundry often evoke memories of Venice. One of them belongs to the church of St Spyridon, the island's patron, who is credited with saving Corfu from plague, famine, cholera and the Turks. His preserved body is paraded round the city on the anniversaries of those miracles to marches oom-pahed by brass bands in full regalia, a legacy of British rule. They wind through streets so narrow the sun barely penetrates.

THE PALACE OF ST MICHAEL AND ST GEORGE, CORFU TOWN

Corfu

This grand building stands at the north end of the Spianada or Esplanade, the second largest town square in Europe. It was constructed between 1818 and 1823, of Malta sandstone, to house the residence of the British High Commissioner as well as the Ionian Senate. It contains two museums, of which the Museum of Asian Art ranks among the world's finest. The statue portrays Sir Frederick Adam, the second High Commissioner, and is but one of many period monuments in this vast space. The British-built Finikas theatre closes off its south end.

KOREA through its landscapes
H KOPEA μέσα από τα τοπία της
南无阿弥陀佛
CORFU MUSEUM of ASIAN ART

THE LISTON, CORFU TOWN

Corfu

These twin arcades, designed by the same architect as those on Paris's rue de Rivoli, are the only intact vestige of the brief French rule. Sitting at a café here, sipping ginger beer and watching a cricket match on the Spianada is a favourite Corfiot pastime. The Old Fort, covering two hillocks opposite, was begun by the Byzantines, finished by the Venetians and remodelled by the British. Other elegant buildings along the square include the graceful Reading Society and the Capodistrias mansion, where Greece's first president was born in 1776.

CAPE DRASTIS

Northwest Corfu

Chalk cliffs soaring above turquoise coves are a recurring motif throughout the Ionian islands. Corfu's north and south coasts possess them, while the west and east shores could not be greener. Paleokastritsa, with its peacock blue waters reflecting tree-covered hills remains a must, and the broad sandy beaches facing the open sea nearby attract hordes of bathers from May through October. Kaiser Wilhelm II, who adored Corfu and built the famous Achilleion Palace near town, had a special 'throne' at Pelekas above Glyfada beach, where he sat to watch the sunset.

AGIOS MATTHEOS

Corfu

Corfu's mountain villages bear no resemblance to the tourist resorts of the coasts. The people of Agios Mattheos, nestled amongst dense greenery in the island's southern half, are farmers. Some of the older women even wear traditional starched white headdresses like those of Belgian nuns. Below the village are several little-known sights: the thirteenth-century Byzantine castle of Gardiki, a grove of olives so twisted and ancient they could illustrate a fairy tale, and Lake Korission, a tree-less lagoon created by the Venetians, where 120 bird species have been counted.

FISHING BOATS AT KASSIOPI

Corfu

Kassiopi, on the northeastern tip of Corfu, was once an active fishing port. Now it's a lively resort, filled with restaurants and bars, but that's history repeating itself. The Romans used to visit often. Tiberius had a villa here and Nero performed at its shrine to Jupiter, though no traces remains. Further down the coast two contemporary VIPs succumbed to its beauty: Sir Jacob Rothschild owns a sylvan estate above Agios Stephanos and Lawrence Durrell composed *Prospero's Cell* while living in the White House on the crescent beach of Kalami.

A COBBLED STREET IN BENITSES

Corfu

Twenty years ago the police dared not enter the lager louts' kingdom of Benitses. Now the charming old village below the Achilleion has been reclaimed and is home to gourmet restaurants and genteel hotels. The Kaiser's Bridge, erected to accommodate his yacht, still juts into the sea nearby. The coast from here to Lefkimmi, the biggest town in southern Corfu, is still green but not as dramatic as the rest of the island. Inland, however, you will find many villages with pastel cottages, flowering balconies and sweeping views of olive-carpeted hills.

GAIOS

Paxos

Paxos, said to have been created when Poseidon chopped off a piece of Corfu with his trident, is tiny (9 sq km/3½ sq miles) and relatively flat. But it still manages to boast three ports, dozens of hamlets and 200,000 olive trees, about a hundred for each of its 2,300 inhabitants. Gaios, the main settlement, is invisible from the sea, hidden in a 'fjord' behind a duo of wooded islets. The faded Governor's mansion, built by the Venetians and inherited by the British, presides over the waterfront. It now houses a folklore museum.

ARIS EXPRESS
N.Π. 159
ARIS

LAKKA

Paxos

Paxos, and its twin Antipaxos, are paradise for boat lovers, whether you have a yacht or a humble dinghy. It is said that the paparazzi trailing Jackie Onassis on her cruises in the Ionian 'discovered' these islands and put them on the map. Even from a taverna set in the middle of Lakka's circular bay you can feast your eyes on the stark limestone cliffs and azure water that make swimming on Paxos so sublime. Exquisite white beaches are a short walk or shorter boat ride away.

LONGOS

Paxos

Laidback Longos is an ideal place to indulge in the Paxiots' favourite form of entertainment, lounging in a café and watching the world creep by. Everyone knows everyone else and no one is ever in a rush. You might have to get up from your seat to let the island's one bus squeeze through the minuscule port but, once the flurry is over, things settle back down to normal and you can make friends with the new arrivals. Deciding which gourmet restaurant to patronize will be your only dilemma.

EVENING LIGHT THROUGH OLIVE TREES

Paxos

On Paxos the olive trees grow so thickly, the earth seems in permanent shade. Paxiots care lovingly for their trees, clearing the ubiquitous rocks to eke out every bit of fertile soil. Scattered among the groves are venerable oil presses and butter-coloured chapels, as well as endless low stone walls. The Venetians were responsible for turning Corfu and Paxos into immense olive groves; the Doge offered 12 gold pieces for every hundred trees planted and then profited from the islands' tremendous oil production, the liquid gold of the times.

NIDRI AND BAY

Lefkada

Lefkada, the yachting centre of the Ionian, barely qualifies as an island; a narrow channel separates it from the mainland. Nidri, halfway down the east coast, has an ideal harbour, protected by a host of islets: Skorpios, where Onassis entertained so many celebrities in the 1960s and 1970s; Madouri, owned by the poet Valaoritis's family, sporting a single mansion; and Meganisi ('Big Island'), the largest of the group. Though Nidri has become for hire offer a splendid escape to tree-lined coves.

WINDOW DETAIL, VASSILIKI

Lefkada

If you could look through this window in summer, you would see a long, practically enclosed bay being carved up by dozens of windsurfers. Vassiliki, in southern Lefkada, is the Ionian's mecca for that colourful sport. Continuing past it, down the finger-like peninsula to the west, you come to Cape Doukato, aka Sappho's Leap, where the poet was said to have ended her life after an unrequited affair (with a man). From here you have stunning views of more islands – Ithaka, Kalamos, Kefallonia – linked by ferry with Vassiliki and Nidri.

THE LITTLE PORT OF VATHI

Meganisi

Though Meganisi, a fishhook-shaped island east of Lefkada, is just 20 minutes from Nidri by ferry, it is a different world. Quiet reigns in its three traditional villages, which seem decades removed from the twenty-first century. Fishermen mend their crimson nets on the dock, while well-fed cats purr next to them. Olive trees grow down to the very edge of pebbly beaches on the gentle north side, but take an excursion to the south and you will enter caves gaping beneath towering white cliffs and dive into bottomless turquoise pools.

VIEW FROM NEAR EXOGI, TOWARDS PLATYTHRIAS

Ithaka

When you come to Ithaka, you understand why Odysseus was so homesick. The island's two steep mountain ridges have so little fertile soil that most of its inhabitants have been forced to wander for a living. But its rugged landscape, extraordinary views – from villages named Anogi ('Above the World') and Exogi ('Out of this World') – and intelligent, hospitable people make it unforgettable. A sign at the main port proclaims 'Every traveller is a citizen of Ithaka'. No ruins have ever been found of Odysseus's palace, but Homeric lore is a cottage industry here.

OLIVE TREE AND CYPRESSES

Ithaka

Wherever you walk in Ithaka – and it is a hiker's delight – you stumble on to place names from the *Odyssey* – Laertes' pig farm, Arethusa's fountain, the Cave of the Nymphs – that add to their already considerable charm. Furthermore, the locals keep up with research, preferring to attend a lecture than tend their (tasteful) boutiques. The package tourist is unknown on this idiosyncratic island, favoured by discerning regulars, Greek and foreign. One was Lord Byron, who swam out to the Lazaretto in the middle of Vathi bay.

KIONI

Ithaka

Arguably the prettiest village on Ithaka, Kioni's nineteenth-century houses ring a deep bay on the northeast coast. And although the secret is out, new buildings keep discreetly to the old style. Three ruined windmills add a picturesque note from above, while paths lead to half a dozen of the island's best beaches. It is also a yacht haven. No wonder that the same people keep coming back year after year. And if they want nightlife, there's Vathi, the welcoming main port on an even deeper bay.

TOM

ESPRESSO

BOATS IN FISKARDO

Kefallonia

Kefallonia, the largest of the Ionian islands, claims the group's highest mountain (Ainos), its longest bay (near Argostoli), unique geological phenomena (caves, underground lakes and more) and a famously eccentric, witty population. In 1953, the island and its neighbours, Ithaka and Zakynthos, were changed forever by a devastating earthquake that levelled almost all their buildings. Only Kefallonia's northern settlements survived intact. Since then the charming port of Fiskardo and the tiny village of Assos, crowned by a Venetian castle, are living reminders of what vanished. Fiskardo has become the Ionian Portofino.

THE MONASTERY OF AGIOS GERASIMOS

Kefallonia

St Gerasimos, who lived in the sixteenth century, is the patron of Kefallonia. Meet a native son and chances are he will bear that name. The monastery attracts thousands of pilgrims every year, but smaller churches may have more appeal for foreigners. Ionian interiors are celebrated for their intricately carved and gilded icon screens and Renaissance-style paintings. From the monastery it is a short drive to the Ainos National Park, known for its endemic black pine woods, shaded trails and stupendous views. The forest suffered when the Venetians logged the pines for masts.

THE LIGHTHOUSE AT ARGOSTOLI

Kefallonia

Originally built in the 1820s by Charles Napier, the British governor, this (restored) lighthouse was one of many improvements made to the island's infrastructure. He also laid the first real roads and the bridge across the Argostoli lagoon. Although Argostoli, Kefallonia's capital, was largely destroyed by the earthquake, the exhibits in its Corgialenios Historical and Cultural Museum poignantly illustrate just how sophisticated, prosperous and cultivated a society it had. Whole rooms have been replicated with mannequins in the latest Paris fashions, lace christening outfits, elegant furniture and other nineteenth-century memorabilia.

THE BELL TOWER AND CHURCH OF AGIOS DIONYSIOS

Zakynthos

The Venetians called Zakynthos 'Zante, fiore del Levante', the flower of the East. It was gentle, green with vines producing currants that fetched a high price in trade with England, and its port town surpassed Dubrovnik in loveliness. The earthquake of 1953 and subsequent fire erased that image, but the Zakynthians refused to accept their fate. They rebuilt their capital exactly as it was, using the same honey-coloured stones to recreate their churches. Only three buildings survived. One was the bell tower and church of Agios Dionysios, the island's patron saint.

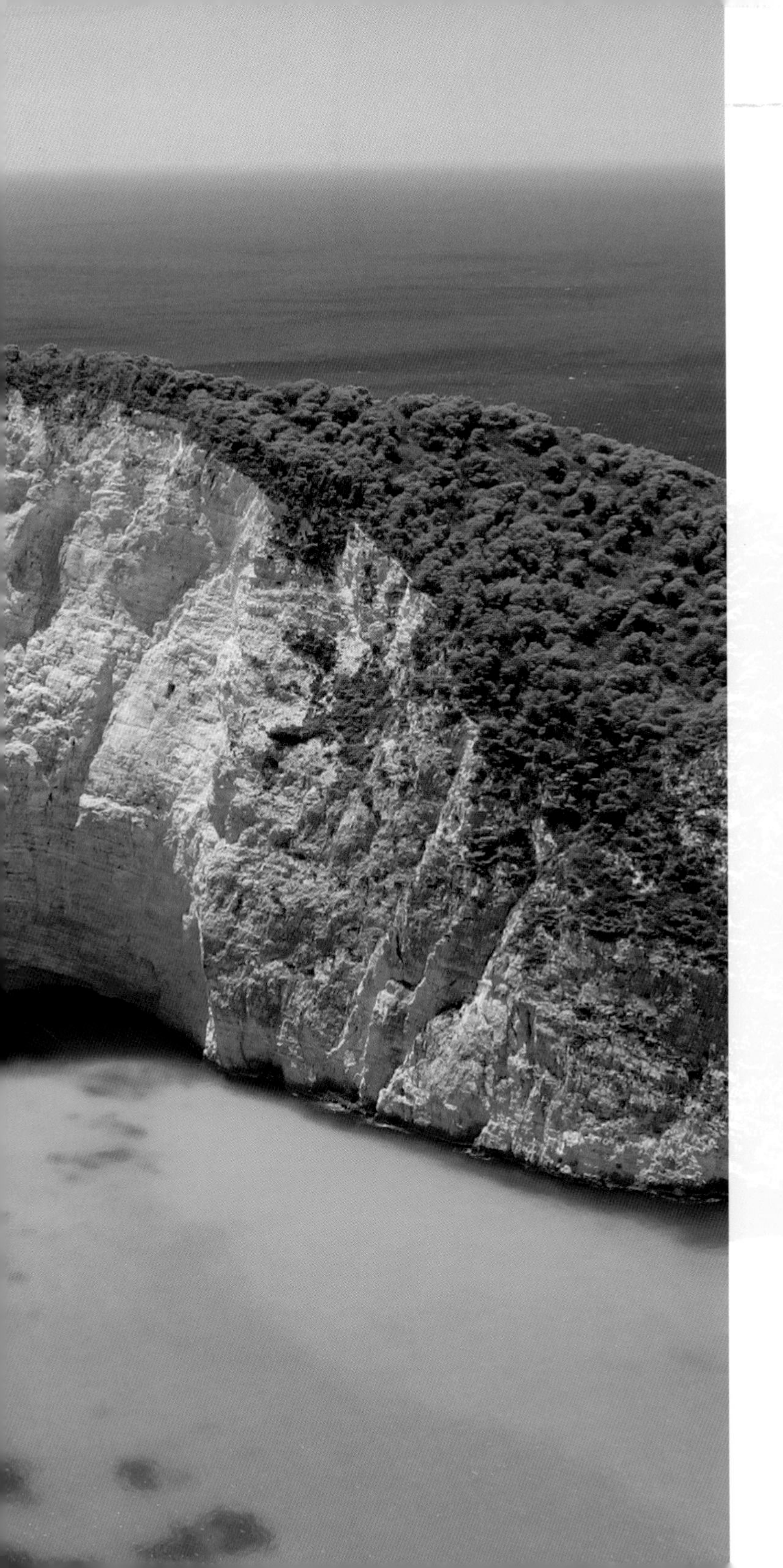

SHIPWRECK BAY

Like all the Ionian islands, Zakynthos has its share of spectacular cliffs rimmed by white sands and aquamarine waters. In summer, the coves fill up with excursion boats. But it also has numerous long, shallow beaches that were nesting places for the endangered loggerhead turtle, *Caretta caretta*, long before tourists and hotel developers discovered them. Today they maintain an uneasy symbiosis, while ecologists campaign against bright lights and sharp umbrella poles. Ironically, the turtles have become tourist attractions themselves, as famous as the bars and clubs at Laganas and Gerakas or the magnificent Shipwreck Bay. This beach gets its name from a boat that was allegedly smuggling cigarettes and wrecked in the early 1980s. Set on dazzling white sand and accessible only by boat, the bay has become an icon of Zakynthos.

THE BLUE CAVES

Zakynthos

The spectacular range of azures, blues and greens in the Blue Caves, in the north of Zakynthos, attracts boatloads of tourists daily in summer. And when you see how much beauty likes in and around the island, it is easy to understand why Zakynthos has produced so many poets. The most famous is Dionysios Solomos, considered the father of Greek poetry. He composed the lyrics for the country's National Anthem, 'Hymn to Liberty', on Strani Hill above the port. The main square bears his name and a statue of Liberty, while the more intimate St Mark's Square boasts the gleaming white Solomos Museum. Andreas Calvos and Ugo Foscolo, who wrote in Italian, also belong to the Zakynthos poets' roster.

AERIAL VIEW OF KAPSALI BAY

Kythera

This bay with its two crescent beaches is what you see from the Venetian castle at Hora above it. Kythera has much to explore: two smaller Venetian forts, monasteries perched on craggy mountains, thousand-year-old chapels, a half-finished British-built aqueduct, gorgeous beaches, a deserted medieval town, Minoan peak sanctuaries, unspoilt villages and even a sparkling waterfall. Hard to get to, difficult to leave, Kythera attracts many foreigners in search of a mid-life change. Their artistic pursuits combine with still vibrant local traditions to create a uniquely pleasant atmosphere in Aphrodite's mythical birthplace.

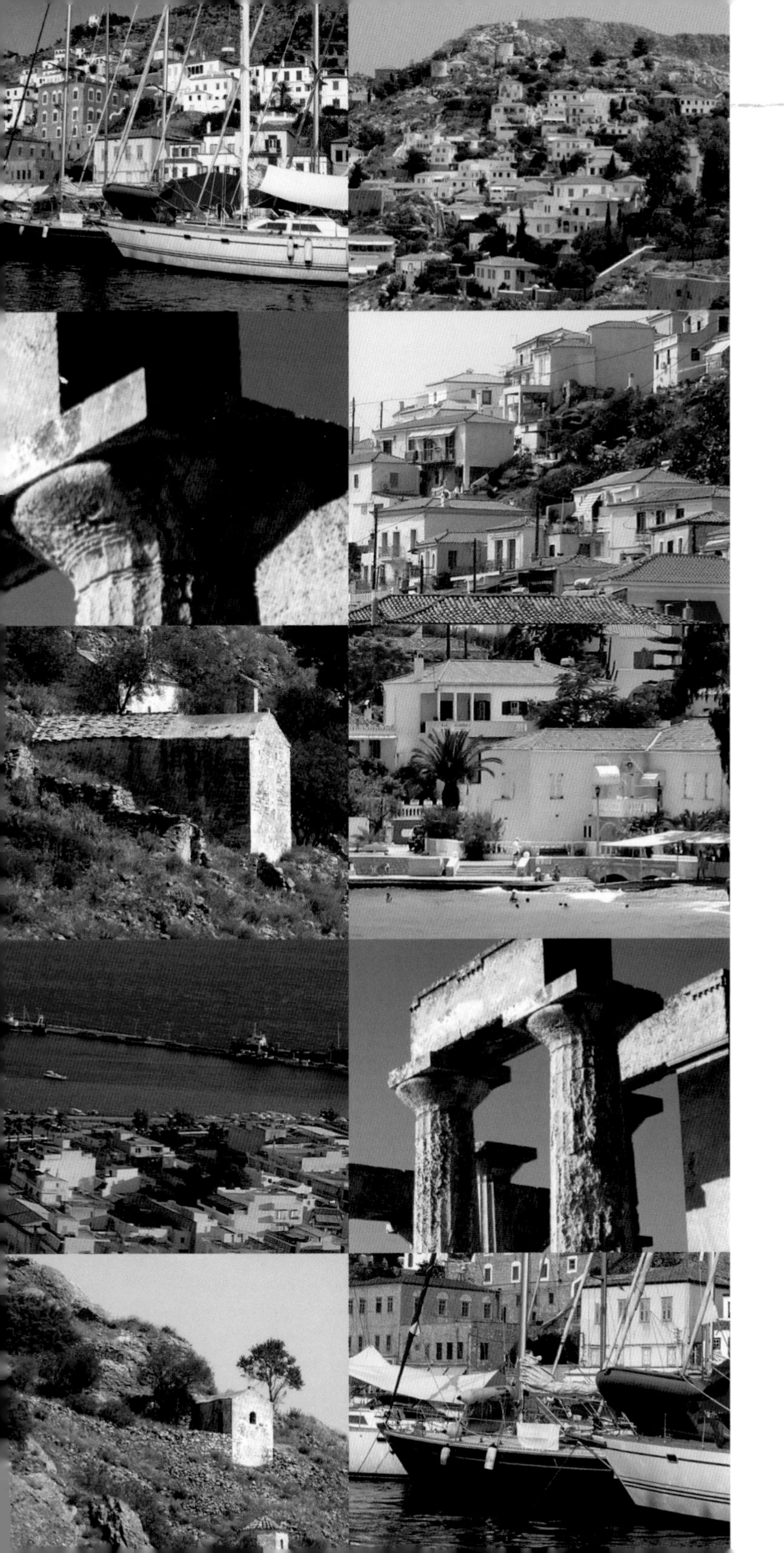

THE ARGO-SARONIC ISLANDS

These are the islands closest to Athens, brought so close in fact by ferries and skimming hydrofoils that Aigina and Salamis have become suburbs, while Poros and Hydra are day-trip destinations. Even Spetses, at two hours and a bit from Piraeus, could be explored in a day, but all deserve more time.

Despite their proximity to the metropolis, each island has preserved its own individual character, even as development encroaches on traditional styles and pristine beaches. Aigina, a city state that challenged Athenian supremacy on the seas, also served as the capital-in-waiting before Greece was fully liberated from the Turks (1828–30). Its temple of Aphaia predates the Parthenon and it still boasts many charming neoclassical buildings from the latter period as well as pistachio groves and pleasant beaches.

Poros, a stone's throw from the Peloponnese, could be a good base from which to visit Epidaurus, Mycenae and Nafplio, while Hydra and Spetses still retain some of the aristocratic airs they possessed as ship-owning islands that provided the armada in the fight for independence. Athenians flock to them in summer and at weekends, and holiday homes sometimes outnumber locals' residences, but regulations curb most excesses and the swimming is still excellent.

THE SHIPYARDS

Salamis

Salamis has a permanent place in Greek and, indeed, Western European history as the site where the Persian navy was crushed on 22 September 480 BC. Themistocles had correctly interpreted the Delphic Oracle's warning to rely on Athens' 'wooden walls' to mean 'invest in ships'. Xerxes watched from Mt Aigaleo across the straits as wily Greek captains outmanoeuvred the cumbersome Persian vessels. That day spelt the end of Asian designs on Europe until the Ottoman Empire two thousand years later. Modern Salamis has a naval base and shipyards but no antiquities to speak of.

TEMPLE OF APHAIA

Aigina

Aigina, in contrast, is full of ruins, most notably this remarkable Late Archaic temple built around 580 BC. Though only 20 of the original 70 Doric columns still stand, it demonstrates the early wealth of the island state as well as the ancients' genius at choosing powerful locations. Set on a pine-covered hilltop, it has a commanding view of the Attica coastline from Piraeus to Sounion. Other antiquities worth a look are the single-columned temple of Apollo in town and an imposing altar to Zeus up Aigina's cone-shaped mountain.

THE MEDIEVAL CAPITAL AT PALIOHORA

Aigina

Once 365 churches occupied this hillside, where the islanders moved to escape pirate attacks in 896. Clambering in search of faded frescoes and Byzantine reliefs is its own reward. Only a few chapels have been restored. In contrast, a massive domed church that can hold 9,000 worshippers stands minutes away. It was built in 1994 to honour Nektarios, the only twentieth-century saint in the Orthodox canon, who lived in a humble cell next door. But do not leave Aigina without sipping an ouzo at the fish market or sampling its trademark pistachios.

VILLAGE OF LIMENARIA

Angistri

Angistri is where people go when Aigina becomes too crowded. Small boats link the two several times a day. It has three villages – Milos, Skala and Limenaria, one bus for its 700 inhabitants and uncounted fishing boats. Summer bungalows are inserting themselves among the pine woods – Eastern Europeans have discovered how reasonably priced it is – but you might visit just for the views of the other islands and the mainland. Angistri, which means 'fish hook', is equidistant from Epidaurus, Methana, Corinth and Aigina. Despite the incursions, it stubbornly holds on to its traditional ways.

THE WATERFRONT

Poros

Visitors to Poros get two for the price of one: an evergreen island with all the concomitant pleasures and a second set of waterfront cafés, beaches and interesting sites on the Peloponnese across the narrow (360 m/390 yd) channel. A calm lagoon shelters yachts in any storm, a tiny islet with a tiny castle half blocks the exit, and one never tires of watching the busy passage. From Galata you can take a donkey ride in the fragrant Lemon Forest or picnic at the waterfall near ancient Troezen, where Theseus was born.

YACHTS IN THE HARBOUR

Hydra

Greece's Gibraltar, the island of Hydra is a precipitous sliver of naked rock 65 km (35 nautical miles) from Piraeus. In the seventeenth and eighteenth centuries, its sailing ships plied the Mediterranean making huge profits from trade that enabled their owners to build grand mansions surrounding the port. So rich was Hydra that it managed to support a population of 35,000. The steam engine snuffed that prosperity but the grandeur remained. Hydra was one of the very first islands to be 'discovered' by foreigners in the late 1950s. Many were artists and poets.

KAMINI VIEWED FROM THE SEA

Hydra

No one visits Hydra for the swimming; there is only one sandy beach. It is an architectural marvel, an ever fascinating composition of imposing stone houses that ascend the stone mountain in impossibly vertical succession, almost to its very ridge. There, four monasteries survey the horizon from wooded copses. In contrast to this austerity is Hydra's image as a party island. Elegant soirees take place behind those high walls, the quay's cafés buzz with gossip, yachts tie up three deep in the small horseshoe-shaped port, celebrities mingle unnoticed by blasé locals.

DONKEY

Hydra

And everyone gets around on foot, while donkeys take the place of delivery vans. Hydra has no roads, no wheeled traffic. Its alleyways are stepped. This is no island for the weak of limb, though the less energetic rely on sea taxis to speed them to Kamini, Vlichos or isolated coves behind the island. They are no help, though, in hopping from bar to bar or choosing a chic taverna tucked in a vine-covered courtyard or converted stately home. Day-trippers just have to imagine Hydra's sophisticated, exuberant all-night life.

AGIOS MAMAS, BETWEEN DAPPIA AND THE OLD HARBOUR

Spetses

Another island that enjoyed prosperity in the age of sail, Spetses too has many great houses strung along its waterfront. But this is a gentle, green island, favoured by Athenian gentry rather than bohemian intellectuals. They gather by day in the Dappia, the minuscule main port with its pebble mosaics and cannon left from the Revolution. At twilight, they head for the Old Harbour, an enclosed bay ringed by gracious homes and restaurants. Kaikis and sea taxis ferry people to beaches round the island or across on the nearby Peloponnese.

NINETEENTH-CENTURY MANSIONS IN SPRING

Spetses

These homes rest on massive cisterns, the island's only water supply in the past. Today, a tanker arrives daily with water from a mainland spring. Nevertheless, flowers thrive in its mild climate, jasmine perfumes the night air and bougainvilleas spatter white walls with every shade of red, pink and yellow. Spetses has two museums (one of which belonged to its famous lady Admiral Bouboulina), the splendid eighteenth-century monastery of St Nicholas, a dozen horse-drawn buggies, five taxis and a few thousand motorbikes. Resisting change, two old-fashioned boatyards still produce wooden kaikis.

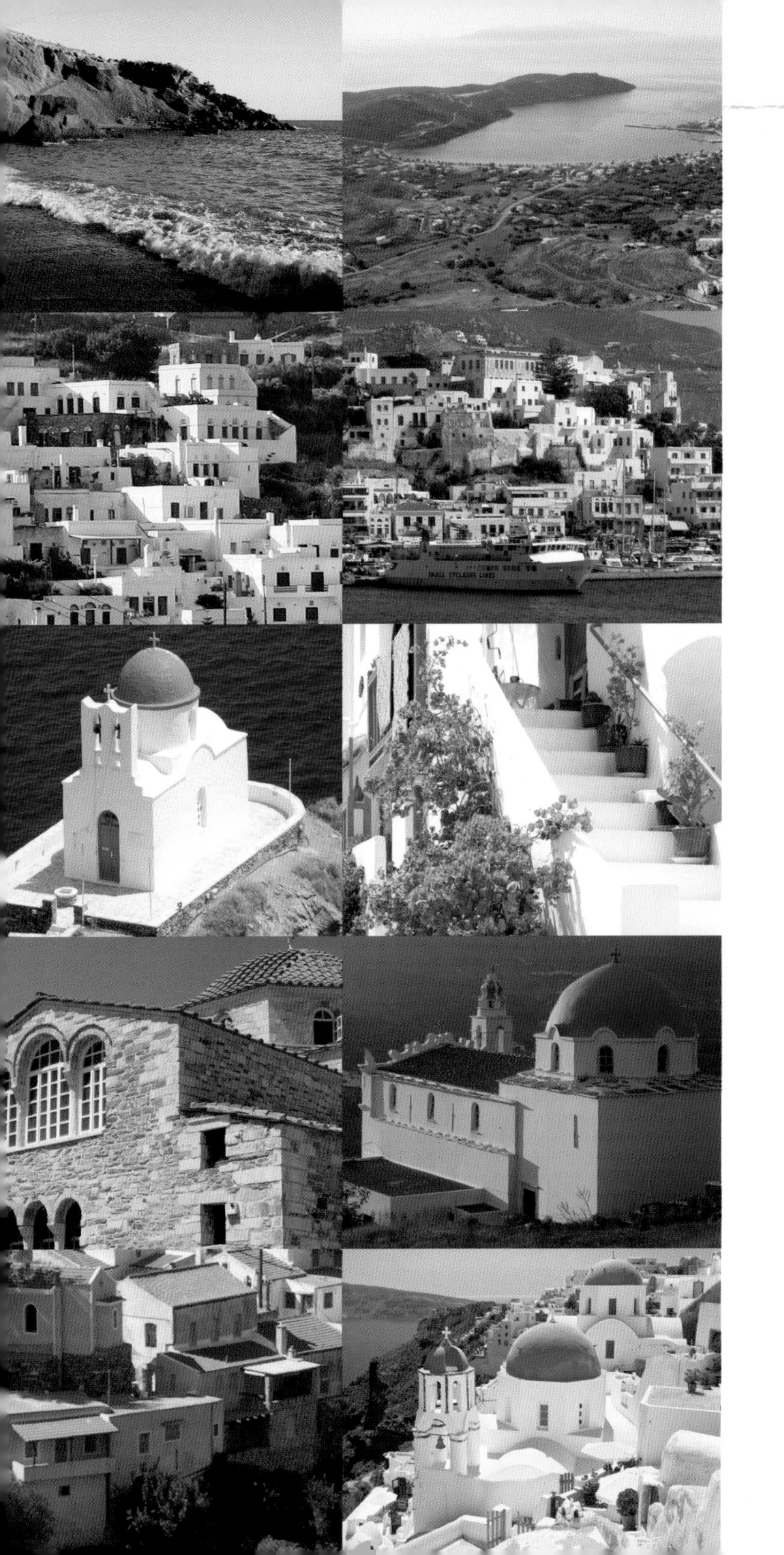

THE CYCLADES

For many people, the Cyclades represent the quintessential Greek island: small sugar-cube houses spilling like clotted cream down a tawny slope, interrupted only by blue church domes that match the enveloping 'wine-dark' sea and sky. The Greeks, feeling the same, chose blue and white as the colours of their flag.

These islands take their name from the (very loose) circle they form around the sacred island of Delos, the birthplace of Apollo. Some of them – Milos, Amorgos, Paros, Naxos – participated in the Aegean's earliest civilization, the Cycladic (3000–1100 BC), famous for its angular statuettes of Parian marble. Santorini flourished during the Minoan era, until the volcano blew it apart (in the fifteenth century BC). Mycenaeans, Athenians, Romans, Byzantines, Crusading Franks, Venetians, Turks – all left marks of varying intensity on these islands.

Despite the common history and cultural similarities so apparent in their architecture, dances, cuisine, vegetation and love of flowers, each one, large and tiny, has its own definite personality. Mass tourism, fast boats and airports, and excessive, tasteless building are making it more and more difficult for some to preserve this character. But seek and you will find it, especially off season, even on Mykonos.

PIDIMA TIS GRIAS BEACH

Andros

Until recently, Andros was a well-kept secret. Few people visited its incomparable beaches, Riviera-like hillsides and gushing springs, although only two hours separate it from the mainland. This northernmost of the Cyclades, anciently known as Hydrousa, or Watery, kept to itself, the preserve of world-class, very discreet shipowners. In time, though, the word got out, chiefly because the annual exhibition at the Goulandris Museum of Modern Art became a summer must. Crossing the island to admire works by Picasso, Matisse or Giacometti, visitors could not help noticing its striking beauty and returning.

VIEW OF BATSI

Andros

Most visitors stay at Batsi, a fishing village that has quadrupled in size in the past two decades. It has most of the island's restaurants, hotels and nightlife, which cannot be compared to the 'scene' at more 'popular' destinations. Instead, Andros draws outdoor types: people who would gladly suffer rutted roads to get to a spectacular cove, who would rather hike along a river lined with watermills than bask on a noisy beach, or who would relish discussion with the abbot of a monastery founded in the 960s. Andros pleases the adventurous.

XOBOURGO AND TYPICAL VILLAGE

Tinos

Religious Greeks make pilgrimages to Tinos hoping for answered prayers. The icon of the Virgin Megalohari in the marble-faced Evangelistria Church in the main port has a miracle-working reputation on a par with Lourdes. On the Virgin's feast days of the Annunciation and Assumption, the town fills to bursting with supplicants, priests, politicians and salesmen hawking candles, devotional souvenirs and Turkish delight. Beyond the hubbub is another world – of compact, white villages, bucolic pastures, weird rock formations and, as atop this peak, a crumbling Venetian fortress on ancient foundations.

IMPOSING CHURCH

Tinos

There is no shortage of churches either, 750 at last count. Surprisingly, a good portion of them are Catholic, a legacy of Venetian domination that lasted until 1715. The Venetians also bequeathed a thousand or so dovecotes, white-washed towers decorated with lacy geometric designs and symbols, found only on Tinos and Andros. In a similar vein, every house possesses openwork marble fanlights above its doors and windows. Whether folk art or fine art, Tinos has bred some of Greece's most admired painters and sculptors, including Halepas, Gyzis and Pheidias's father.

AN OLD WOMAN WALKS UP TO THE SQUARE

Tinos

Paved or cobbled, paths of all kinds crisscross the island, making it sheer delight for hikers. They may lead to a village square with its church, fountain house, sculptor's workshop and small taverna. Tiniots cook with finesse and imagination. You could also find your way to Poseidon's shrine, placid coves or windswept beaches. Windswept is more likely. According to myth, wind god Aeolos made his home on Tinos. Despite constant buffeting, the islanders are cheery and hospitable. Talk to them and buy their capers and cheeses at the farmers' market in town.

IOULIS OR HORA

Kea

Kea, also called Tzia, lies just one hour from Lavrio, a large but little-used port near Sounion. Well off the tourist track, it is rapidly becoming a favourite home away from home for well-off, 'intellectual' Athenians. Ioulis, or Hora, has been continuously inhabited since the seventh century BC, when Kea boasted four important cities. Cycladic-era ruins, a smiling Archaic lion, grey-stone houses, a dearth of hotels and the prevalence of dark oak forests over olive groves give Kea a colour quite distinct from the rest of the archipelago.

ERMOUPOLI

Syros

Nowadays the Cyclades revolve around Syros, not Delos. Ermoupoli is their administrative capital, but its heyday lasted from 1830 to 1880, when more ships docked here than at Piraeus, and it provided 94 per cent of Greece's GNP. Founded by refugees from Chios, who named it after Hermes, the god of commerce, it has a monumental town hall, an opera house modelled on La Scala and super-elegant neoclassical homes. But an older town overlooks it. Catholic Ano Syros dates from the thirteenth-century Frankish occupation and its filigree alleys lead to Renaissance churches.

DELPHINI BAY NEAR KINI IN SPRING

Syros

If southern Syros claims the most sumptuous town in the Cyclades, the northern half of the island competes for some of its wildest scenery. There amidst the wind-blasted crags archaeologists unearthed 6,000-year-old Cycladic ruins. Why not in the tame and fertile south? That is where modern settlers built grandiose mansions in flush times and scores of 'bungalotels' have sprung up lately. But knowledgeable visitors come for the food not the beaches. San Michali cheese, wind-dried fish, caper salad, spicy sausages and, most famously, *loukoumia* (Turkish delight) are among the temptations.

MERIHAS

Kythnos

Kythnos, between Kea and Serifos, rarely figures in foreigners' travel plans. It lacks both glamour and drama, but nowadays that could be a plus. Greeks enjoy it for family holidays, while in earlier years they used to travel there to take the waters at the hot springs at Loutro. Once the country's finest spa, the bath house is a period jewel, built in 1860 by King Otto and Queen Amalia, with royal-sized, marble tubs. The disdainful locals find it useful for laundering *flokatis* (wool rugs). Merihas, the port, has less obvious appeal.

HOUSES AT DRYOPIDA

Kythnos

Away from the sea, Dryopida, the medieval capital, is also the island's prettiest village. It is a pleasant walk from Merihas in summer, especially when smiling housewives emerge from their courtyards with platters of fresh figs, pitchers of cool water and an excuse to pause. But for real refreshment, do not miss a swim at Kolona. There, a strip of golden sand connects Kythnos with the islet-chapel of Agios Loukas and its one ancient column. You can bathe off either side in turquoise water. Naturally, yachtsmen have discovered it is a magical anchorage.

LITTLE VENICE AT SUNSET

Mykonos

Early tourists had to pass through Mykonos in order to inspect the antiquities on nearby Delos. Rumours soon circulated that the present held more charms than the past. In a twinkling, thanks to its incomparable architecture, superb beaches and engaging people, Mykonos became the new 'in' destination for jet-setters, followed by gays, partying yuppies and cruiseboat passengers. But its fragile beauty is being compromised by too many visitors and too much building. Here in Little Venice, a front row seat for viewing the sunset in a former sea captain's house can be pricey.

TYPICAL CYCLADIC ALLEYWAY

Mykonos

Some say that the web of alleys linking Mykonos's close-knit houses was intended to confuse pirates. Others maintain that its seamen were pirates themselves, using their water-lapped houses at Little Venice to unload booty. Lending credence to this theory is the fact that most other island capitals are hidden in the interior far from roving marauders. Whatever their origins, these narrow lanes could not be more attractive. Whether lined with boutiques and bars or neat, colourful houses like these, they are immaculate; even the paving stones are whitewashed every spring.

PRIVATE HOUSE
KEEP OUT

WINDMILLS

Mykonos

Windmills, a Mykonos trademark, no longer grind wheat but still beguile photographers. This row of four stands above a mellow fort built by Venetian overlords in the thirteenth century. From here, you can see some of the island's much-vaunted southern beaches. The whole coast is undulated like ruffles on a curtain, with one white strand after another, interrupted by low rocky promontories. The most famous were baptized Paradise and Superparadise, but guidebooks rate them lower than the more prosaic sounding Elia, Agrari and Lia, which truly conjure up tropical bliss – off season.

PETROS THE PELICAN

Mykonos

When and why did Petros the Pelican become the mascot of Mykonos? One thing is for certain, this is no longer the original bird. Furthermore, two of the ungainly creatures now waddle among the boats and tourists. They must be among the most photographed feathered personalities alive. You can see that Cycladic kaikis bear the same colours as the wooden trim on house balconies, staircases and shutters. Turquoise is common; it offers protection against the evil eye. In the background, one of the island's 500 churches watches over the old port.

PARAPORTIANI CHURCH

Mykonos

Of all the churches, this is the best known. It looks more sculpted than built and this one building actually comprises five chapels, four moulded around the central, domed core. They were added during the course of construction, which began in 1425 and lasted more than 200 years. The church, which is dedicated to the Virgin, takes its name from 'porta', door in Greek, since it guards an opening in the medieval wall that surrounded the town. Centuries of whitewash have softened all edges, rounding once sharp angles into a poem.

TEMPLE COLUMNS

Delos

Delos ranks with the Acropolis, Olympia, Delphi and Knossos as one of Greece's top archaeological sites. In some ways, it is the most rewarding because you can wander freely among the sanctuaries with their temples, but also through streets flanked by one- and two-storey houses. Located around the theatre, several still have wonderfully preserved mosaic floors. Apart from these, the island's most celebrated feature is the magnificent terrace of the lions. Five of the original nine can be seen in the site's museum, along with artefacts spanning over 2,000 years.

VIEW OF LIVADI FROM HORA

Serifos

Serifos may just have the most visually exciting hora in the Cyclades. From a distance, its white houses dribble like melted vanilla ice cream down the side of a giant cone. From within the village, perched on the island's highest mountain, you have dizzying views of beaches scalloping the coast. The islanders moved up here in the early Middle Ages, hoping to thwart pirates. An eyrie of a Venetian castle looms above their houses, now transformed into the epitome of Greek style. Livadi, the port, has its own understated appeal.

LIA BEACH

Serifos

The beach at Livadi is clean enough to swim off, except in August, and there are plenty of others to choose from. Unlike Mykonos, most Serifos beaches are still devoid of umbrellas, lifeguards and refreshment stands serving music with their cheese pies. At one, Megalo Livadi, rusted mining installations are a poignant reminder of the island's industrial past. Exploited in ancient and Venetian times, the iron mines were reopened after Independence by a German company but abysmal conditions led to strikes in 1916 in which four workers were killed.

SEVEN MARTYRS CHURCH

Sifnos

Sifnos next door was another wealthy island, but its silver and gold mines gave out by the fifth century BC. If Serifos is rugged and masculine, Sifnos has a gentle, more feminine character. Its twin horas, Apollonia and Artemonas, invisible from the sea, sprawl over broad hills sprinkled with olive trees. Dry-stone walls slice the land into cultivable strips as regular as if drawn with a ruler. Many a blue-domed church squats amongst them and everywhere else you look – down gullies, on sea-battered rocks, mountain peaks, beaches ...

PANAGIA CHRYSSOPIGI CHURCH

Sifnos

Some churches are stone, but most are white; all are photogenic and no two are the same. The white surfaces glisten like new meringue, fashioned by a master pastry chef into imaginative confections. The art lies in the details: the swelling apses, cupolas, bell towers, barrel vaults, planes and angles, which come together as abstract marvels. Some feature ornamental pottery, an island speciality. Many potters still use the wheel, turning out unglazed and painted crockery and figurines. Clay pot casseroles are delectable in Sifnos, from where Greece's best cooks hail.

A POPPY FIELD

Sifnos

Is there anything more beautiful than Greece in spring? This poppy field lies just outside Kastro, where Sifniots sought refuge from pirates from the thirteenth century until 1836. Built by the Venetians on an unscalable cliff above the sea, this was first the site of the ancient acropolis. Its houses share windowless exterior walls that render them even more fortress-like, while the interior has only two main streets that often resemble tunnels. Stray columns, sarcophagi and Frankish escutcheons lend romance, but just 40 souls are hardy enough to winter here.

THE WATERFRONT

Antiparos

Thirty years ago Antiparos barely registered on any horizons. Ten minutes from Paros, it was a sleepy, one-hamlet island that a person might visit to see its spectacular cave, its only claim to fame. The immense stalagmite at its entrance is reputedly 45 million years old – Europe's most venerable. In 1673 the French ambassador to Constantinople interrupted his search for antiquities to celebrate Christmas mass there. Nowadays the proliferation of luxury villas shows that Antiparos has been discovered by the Athenian gentry. It also calls itself the octopus capital of Greece.

NAOUSSA BAY AT SUNSET

Paros

Some tourists call Paros 'the poor man's Mykonos' because of its reputation as a party island and more moderate prices. Its beaches may lack Mykonian allure but they still attract huge crowds; its main port, Parikia, may be the busiest in the Cyclades since it is ideally placed for island hopping; and on the huge bay in the northern part of the island, Paros's other port, Naoussa, also throbs with life. A wee chapel sits amidst fishing kaikis, café tables surround the minuscule harbour. Though it stretches to bursting point in August, popularity has not tarnished its pristine beauty.

LEFKES FROM A DISTANCE

Paros

But it is not all hustle and bustle. Behind the hectic waterfront of Parikia, a typical island village lurks, with those cool alleys, white walls, tumbling bougainvilleas and even a modest remnant of a Frankish fort, studded with ancient column rounds and hefty marble blocks. And drive into the hinterland of Paros and you might think you are on another island. Lefkes ('poplars' in Greek) is not only green, it is quiet. High and tranquil, it was the medieval capital and still feels removed from modern angst, especially when you enter Agia Triada, the splendid church crowning it.

EKATONTAPILIANI CHURCH IN PARIKIA

Paros

Begun during the reign of Constantine the Great (AD 280–337), this must be the oldest church in the Aegean. It is also among the most impressive. The name means 'Our Lady of the Hundred Gates'; they say 99 have been found. It is composed of three distinct churches. The largest, Agios Nikolaos, has a Parian marble icon screen and columns. The exquisite crystalline marble from the now depleted quarries at Marathi, near Lefkes, was the ancient world's finest. It inspired the Hermes of Praxiteles and the Aphrodite (Venus) of Milos.

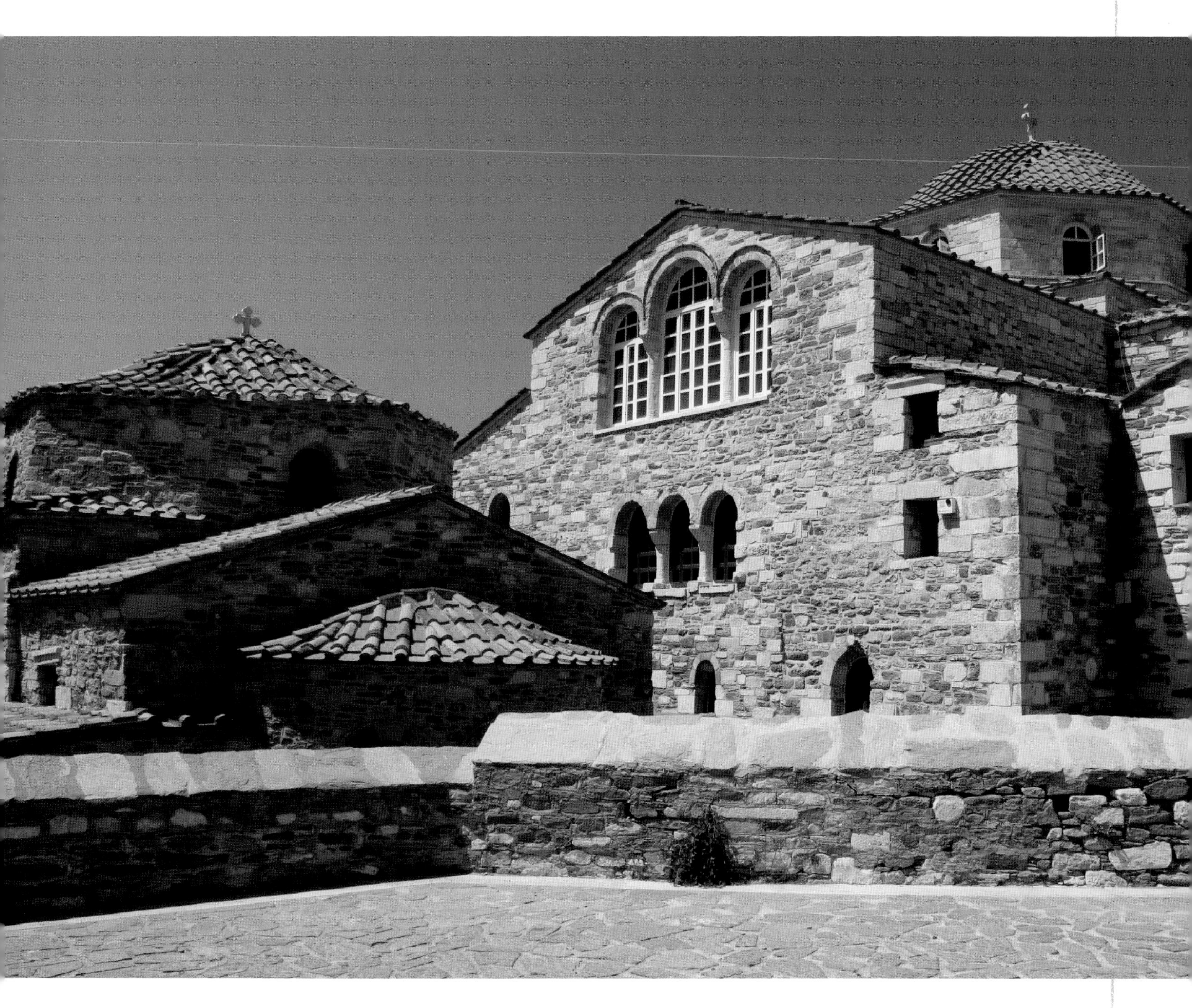

THE PORT AND MAIN TOWN

Naxos

Naxos is the largest of the Cyclades, its Mt Zas (think Zeus) the tallest peak. Fabulously rich in antiquity, Naxians paid for Delos's Lion Terrace; later the Venetians chose it as their headquarters. From here they ruled the islands for more than 300 years. Their fortified tower-houses are scattered throughout the countryside, which ranges from well watered to bone-dry. Trekkers can spend days hiking to sixteenth- and seventeenth-century monasteries, Byzantine churches and marble-paved villages like Apeiranthos and Halki, while sun worshippers will have a hard time deciding which gorgeous beach they prefer.

SMALL CYCLADES LINES

PORTARA, GATEWAY TO NAXOS

Naxos

This enormous door belonged to a temple of Apollo, begun in the sixth century BC but never finished. The symbol of Naxos, it stands on an islet near the port. Supposedly this was where Theseus abandoned Ariadne after defeating the Minotaur, and where Dionysos rescued her. The island has other unique antiquities: an Archaic temple of Demeter in golden stone, the old quarries and two massive marble statues (*kouroi*). The smaller (6.5 m/21 ft) sleeps in an inland vegetable garden; the other (11 m/36 ft) lies amidst rocks on the north coast – puzzles to archaeologists.

COVERED PASSAGEWAY IN KASTRO

Naxos

You cannot see it until you are in it. The waterfront is a rather ordinary mix of souvenir shops, ticket agents and eateries, but step behind them and you will enter the thirteenth-century Venetian citadel. The last of the original 12 towers stands near the north portal. From there, dark lanes wind up to the castle proper, passing half-timbered houses emblazoned with coats of arms. Sharing the top square with the Duke's palace are a Catholic cathedral, two monasteries and a not-to-be-missed museum with a superb collection of Cycladic idols.

PANAGIA THALASSITRA CHURCH IN KASTRO, JUST ABOVE PLAKA

Milos

Since 7000 BC Milos has owed its livelihood to geology. In prehistoric times, obsidian, black volcanic glass from Milos was exploited for arrow heads and sharp tools that were shipped as far as Egypt and Italy. The Romans mined it for bentonite, sulphur, pumice and many other minerals. Today, it is the EU's biggest source of perlite and bentonite. Milos does not need tourists, so the alleyways and tavernas of Plaka are refreshingly calm, pretty and welcoming. But with so many companies plundering the interior, visitors should stick to the coast.

ROCKY COASTLINE

Milos

Like Santorini and Nisyros, the island belongs to the Aegean Volcanic Arc that runs from Corinth to Turkey. Its volcano last erupted 90,000 years ago, but left amazing rock formations (as well as subterranean wealth). Excursion boats visit them all, chugging across the immense natural harbour (among the Med's largest), before passing crimson towers, ebony rods, rocks smooth as yogurt or red as tomato paste, yellow-streaked cliffs, aquamarine pools, cleaver-cut fjords, cathedral-like grottoes and some 70 beaches. At one, boiling water bubbles through the sand. Milos is anything but dull.

HORIO

Kimolos

The excursion boat pauses for lunch at Kimolos on its journey around Milos, but more and more people are heading straight there. With just one inland village, Horio, offering few distractions, this is a place to recover from work, winter, whatever. When revived, you can investigate the inevitable Venetian fortress, an ancient necropolis, a sunken ancient city, a couple of caves, and take a kaiki or a donkey to at least 12 beaches. One has hot springs, another has the famous chalk for which the island is named and still supplies.

THE HORA AND CHURCH OF THE PANAGIA

Folegandros

The Hora of Folegandros, in the southern Cyclades, teeters at the edge of a precipice, 210 m (690 ft) above the crashing sea. The zigzag path branded like white lightning on to the stony hillside ends at the 'cathedral'. Visible from virtually everywhere, the Panagia is the island's emblem. Though old timers talk about development, the hora still has an intimate Sixties atmosphere. Jolly tavernas fill its five interlocking squares, presided over by cosy churches. With its steep slopes laddered by terracing, this is the land of the prickly pear, thornbush and warm smiles.

VIEW OF HORA

Sikinos

Sikinos, next door to Folegandros, is even further off the beaten path. Daily excursions are offered between the two but often cancelled by choppy seas. Just 300 people call it home. Most live in the charming hora, which combines the old Kastro with the newer Horio. Grand old mansions, a fortified monastery, windmills and some abandoned houses add to the romance. The island's main 'sight' is the seventeenth-century monastery of Episkopi built over a seventh-century church resting on foundations of a Roman mausoleum. Donkeys are far from outmoded here.

THE HORA

Ios

Walk around Hora by day if you want to enjoy its typical Cycladic architecture, with traditional churches, labyrinthine lanes, blue-shuttered houses and even two museums. But come in the evening if you would rather cruise the hora's famous bars and discos, which promise everything from tabletop dancing to funk, Latin, rock, reggae and even the classics and jazz. For many decades now, Ios has been the almost exclusive preserve of young backpackers, intent on heavy drinking and all night partying. A festival at the new Elytis amphitheatre offers a more cultural alternative.

CLICK cafe

MYLOPOTAS BEACH

Ios

Long before the bars, there were the beaches. Long expanses of pale sand skirting psychedelic shades of blue (even when sober) are what attracted visitors in the first place. And they still captivate, even when crowded or slashed by jet skiers and windsurfers. Mylopotas and Manganari are the best known, but those in the southeast, furthest from Hora, win the highest kudos. One wonders what Homer, whose alleged tomb lies on Mt Pyrgos, would think of the new Ios. Most probably, he would just pour a libation and join in the fun.

SIGNPOST ON STONY HILLSIDE

Amorgos

Slender and precipitous, wild Amorgos is mostly craggy mountain ridge, crisscrossed with goat paths elevated to hiking trails. It has two very different ports: green(ish) Katapola, near the scant remains of a Minoan city; and Aigiali, a budding resort, on a fine slice of beach. The hora, dubbed 'enchanting' by guidebooks, combines sophisticated boutiques with timeless architecture. The island's prize is the eleventh-century Hozoviotissa monastery, a blindingly white, gravity-defying edifice that clings to the sheer rock near Hora. The cult movie *The Big Blue* lifted Amorgos on to the international tourist scene.

IA

Santorini

Santorini, one of the marvels of the world, rests on shaky foundations. Since it first erupted in the fifteenth century BC, the volcano has boiled over about 14 times, giving birth to new lava islets in the caldera, as well as smoke and chthonic rumblings. Santorini attracts over 80,000 visitors each year, but even too much popularity cannot dim the allure of standing on the edge of the black cliff and gazing on to the white villages rimming the ultramarine bowl. Ia, its loveliest village, has made an industry of sunset-watching.

FIRA

Santorini

Somewhat more commercial than Ia, the capital has the same idiosyncratic architecture – houses and even hotels literally dug into the cliffside. Barrel-vaulted rooms looking on to infinity pools reached only by negotiating dozens of steps deter few guests. Besides the compelling view, Fira has a splendid new museum devoted to finds from the Minoan site at Akrotiri, from bath tubs to jars decorated with swallows and dolphins, and those incredible frescoes showing the saffron gatherers, the blue monkeys, the boy with the fish … Ponder them over some outstanding Santorini wine.

RED BEACH AT AKROTIRI

Santorini

Excavations at Akrotiri, Greece's Pompei, did not begin until 1967. Since then, only 3 per cent of the city has been unearthed, yielding some 10,000 finds dating back to 3000 BC. But we will not see most of them. Archaeologists say 200 years are needed to analyse and treat them before they can be displayed. Console yourself with a swim off Santorini's red and black beaches, a clamber round the huge site of Ancient Thira (seventh century BC, Roman era) and a feast with Santorini tomatoes. Food grown on volcanic soil tastes so good.

DESERTED BEACH

Anafi

The ultimate getaway from civilization, except in August. Even though boat schedules are erratic, the promise of secluded beaches away from the masses is beginning to lure people to this little known, tiny island southeast of Santorini. Anafi's only fame derives from the fact that it supplied the labourers who built the new capital at Athens after 1840. Their own houses still stand in the Anafiotika district of Plaka below the Acropolis. Anafi has all one requires for a quiet holiday: a pretty hora, walks, beaches and a dragon's cave.

CRETE & THE DODECANESE

Like all the big islands in the Mediterranean, Crete has its own history and culture, quite distinct from that of mainland Greece and the rest of the Aegean. It was the birthplace of Europe's first civilization, the Minoan, which has left enduring legends and exquisite art, and has continued to attract invaders ever since, from the Mycenaeans to today's tourists.

Long and narrow, Crete is incredibly diverse with high mountains, at least 2,000 species of plants, dozens of ancient sites, Venetian castles, healthy food to die for, as well as swanky resorts and tacky package 'slums'. Hospitable and prosperous, Crete is a world unto itself.

The Dodecanese ('Twelve Islands') dangle like a string of beads along the coast of Turkey, swinging towards Crete to link Karpathos and Kassos, almost touching Turkey to embrace Kastellorizo. They include the famous – Rhodes, Kos, Kalymnos – and the unknown – Tilos, Lipsi, Agathonisi – and they number far more than 12. Although some – Patmos, Astypalea – bear the white architecture of the Cyclades, most reveal a different heritage, resulting from their proximity to Turkey, occupation by the Knights of St John and the twentieth-century presence of Italians. The Dodecanese were not united with Greece until 1948.

AGIA TRIADA MONASTERY, AKROTIRI

Crete

What better introduction to Crete than a monastery? Near Chania, this grand seventeenth-century edifice sits on the peninsula that begins at Souda Bay, now a NATO base. The style, more Renaissance than Byzantine, is typically Cretan. Monasteries have more than once played a significant role in the island's history. Arkadi, near Rethymno, became a symbol of resistance to the Turks when, under siege in 1866, the abbot blew up the powder stores, himself and 964 fellow Cretans rather than surrender. And in 1941 the monastery of Preveli, on the south coast, hid Allied troops until they could be evacuated.

THE PORT OF CHANIA

Crete

Chania, the most westerly of Crete's cities, is arguably the most attractive. With no visible Minoan remains and just a stretch of evocative wall from ancient Kydonia, it is the Venetians' La Canea that we love. Massive fortifications, the huge shipyards and the imposing lighthouse are all impressive, but behind them the mansions and medieval streets, the atmospheric shops and restaurants hold even more charm. Peppered with mosques and minarets, a fabulous covered market and innumerable spots to sip raki, Chania never ceases to fascinate. It even has its own strictly local customs and foods.

THE SAMARIA GORGE

Crete

Behind Chania loom the often snowcapped White Mountains. Hidden among them is the Omalos plateau and the entry to the Samaria Gorge, one of Europe's longest and most dramatic. It is 16 km (10 miles) long and only 3 m (10 ft) wide at the Iron Gates shown here. The trek begins at 1,250 m (4,000 ft), and it's downhill all the way to the sea at Agia Roumeli. It can get crowded (up to 3,000 hikers a day) but wild flowers, unbelievable scenery and a sense of accomplishment reward all who pass through.

FRANGOKASTELLO

Crete

The coast from Agia Roumeli to Frangokastello is naked and uninviting. Grey-beige cliffs plunge to the sea, interrupted only by two small villages, Loutro and Hora Sfakion, where hikers pick up the bus for Chania. This stark region was always fiercely independent, its men defiant of any authority, whether foreign or Greek. This Venetian fortress was never used, but in 1828 it witnessed a horrific battle between Cretans and Turks. In late spring, the ghosts of the slain fighters are said to return, shades in the mist called Drosoulites.

MINARET, RETHYMNO

Crete

This minaret, Rethymno's tallest, encapsulates the city's history. It belongs to the Nerantze mosque, which began as a Venetian church and now houses the conservatory. Rethymno's history started with Venetian rule in 1204. There Crete's largest fortress was established, and the old town of more than 600 surviving houses contains the best-preserved Renaissance domestic architecture outside Italy. Rethymno makes the most of its heritage. A Turkish jail houses the archaeology museum, and the soap factory a modern art collection. The unrelenting cement on the long beach front, however, is another story.

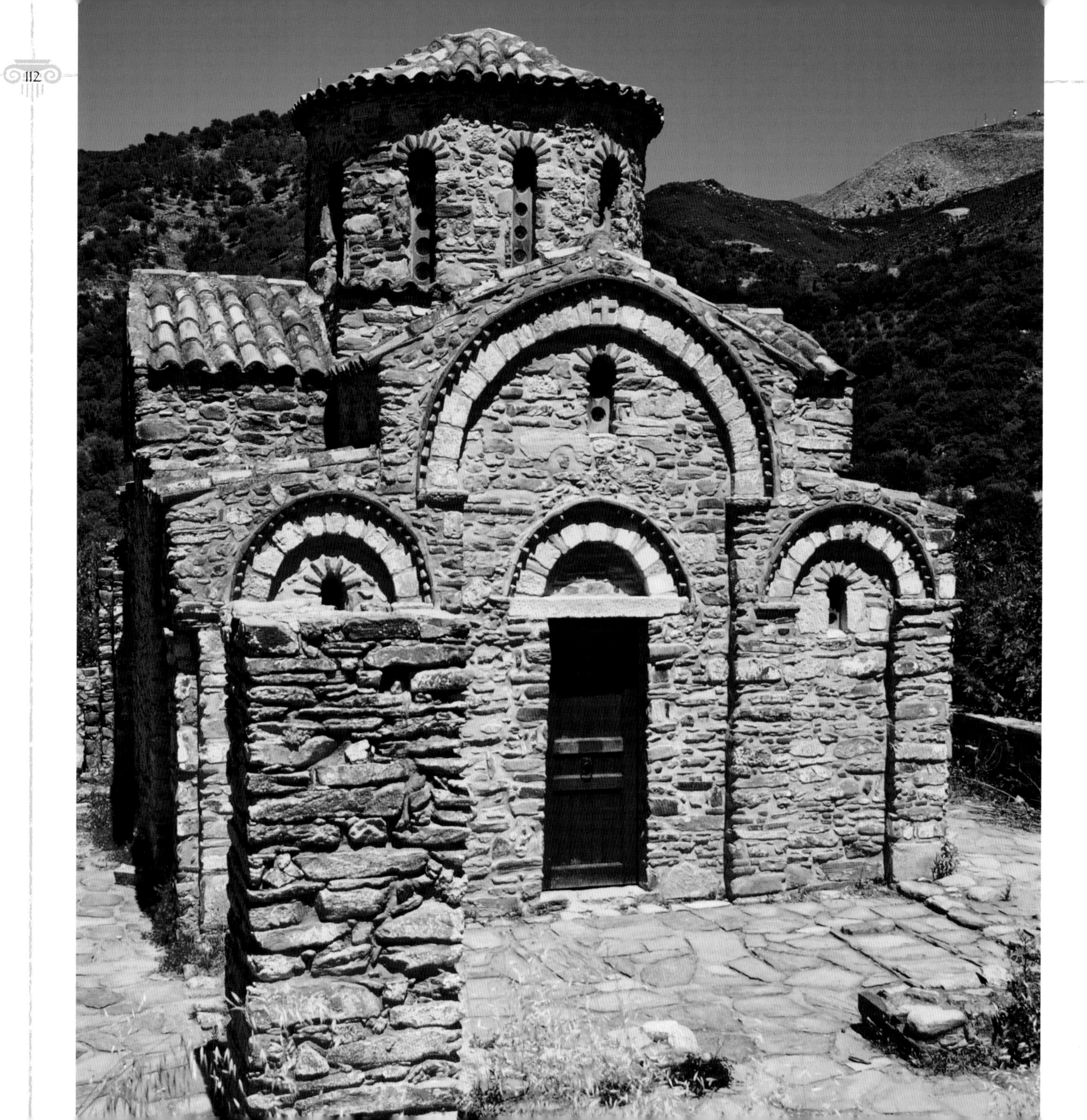

BYZANTINE CHURCH IN FODELE

Crete

This is one of Crete's more than 600 Byzantine churches. A fourteenth-century gem, it stands very near the house in which Domeniko Theotokopoulos, better known as El Greco, was born in 1541. Fodele lies close to Heraklio, Crete's capital. Although it possesses the requisite Venetian port and fortifications, fountains and other period architecture, it also has many constructions in 'late Cretan chaotic' style. More than two million tourists flock to the city annually. Most only visit Knossos and the museum with its incomparable treasures. Lively Heraklio will reward those who dig deeper.

THE MINOAN PALACE OF KNOSSOS

Crete

The most famous, best preserved of all Minoan sites, Knossos covers 20,000 sq m (65,500 sq ft) of labyrinthine chambers, corridors, courtyards, storerooms, staircases and plumbing. They actually comprise the ruins of three palaces, constructed between 2000 and 1400 BC. When Sir Arthur Evans began excavations in the late nineteenth century, his finds seemed to confirm at least some of the ancient myths, while revealing an extraordinarily elegant lifestyle. Though purists criticize his brightly painted reconstructions, they make the ruins more comprehensible. Untouched, the palaces at Phaistos, Mallia and Zakros impress us differently.

AGIOS NIKOLAOS

Crete

Heading east from Heraklio, you come to Agios Nikolaos, which has long outgrown its fishing village origins. It is now one of Crete's more upmarket resort areas. Elounda, a transformed stretch of rocky coast to the north, may boast more luxury hotels than anywhere else in Greece. Contrasts abound. Melancholic Spinalonga was a leper colony until 1953, while life in the traditional hill villages continues oblivious of tourism. The Lassithi plateau is all orchards and farms – another source of Cretan wealth – where visitors buy potatoes, not souvenirs.

PALM GROVES AT VAI

Crete

Is this Crete or some South Sea island? You are unlikely to be alone on this spectacular beach, but to the north smaller coves mimicking it offer more privacy. This is the island's once remote east coast. Its mountains are lower, its monuments more humble, but the friendliness of the locals and relaxed atmosphere linger in the memory. Among its sights are the fortress monastery and organic vineyards of Toplou and the Minoan palace at Kato Zakros, conveniently situated near a pristine black-pebble beach ringed with fabulous tavernas.

MONASTERY OF AGIOS MAMAS

Kassos

From rocky, barren Kassos, this deserted monastery looks south to Egypt. Few tourists veer this far off the beaten track, but they're welcomed when they do. It's hard to believe it was once prosperous but, in the nineteenth century, its population reached 11,000, many of whom were seamen. Some 5,000 Kassiots worked on the Suez Canal. Another 7,000 were massacred by Turks in 1824 – the island has its own Holocaust Day. Most visitors stay at Fry – one hotel, one nightclub, one taxi – with its pretty fishing harbour.

PIGADIA BY NIGHT

Karpathos

Karpathos could be two islands. Its long and extemely rugged mountain chain has blocked communication between north and south so effectively that they have evolved very differently, with the north retaining far more of its historical appearance and traditions. Pigadia, the capital since 1894, presides over the fertile southern half. It suffers from too much concrete but is agreeable nonetheless. From here buses travel to amazing beaches, early Byzantine churches, the 'Cyclopean' walled acropolis of ancient Arkaseia and unspoilt mountain hamlets. No buses connect north and south, just boats. But be warned, the Karpathian Sea is notoriously stormy.

LEFKOS BAY

Karpathos

As you can see, Lefkos, on the middle of the southwest coast, is a budding resort. It boasts not one but five delicious beaches. From here boats ferry sightseers to a medieval fort on Sokastro islet, while hikers can walk inland to Mycenaean tombs. Southern Karpathos, though wild, is surprisingly green, mellowed by pines dotting its mountains, apricot, pomegranate and fig orchards engulfing its hamlets and flowers in bloom year round. At Othos, a museum in a two-roomed house shows how Karpathians lived up until the mid-twentieth century.

OLIMBOS

Karpathos

But to really venture into the past requires a two-hour boat trip north to Diafani and a bus ride to Olimbos, a high, arid village with 70 windmills, four in full sail. Olimbos is a phenomenon in modern Greece. It is a matrilineal society and its women still wear their age-old costume of long black skirt with bright embroidered apron, jacket, headscarf and soft leather boots. Their dialect has some Doric words, their music ancient modes, and house keys resemble those described by Homer. Unfortunately, our curiosity has embellished tradition with kitsch souvenirs.

AGIOS NIKOLAOS FOUNDOUKLI

Rhodes

Rhodes, the queen of the Dodecanese, is more famous for resorts than Byzantine churches, but they do exist. This one, at the centre of the island, dates from the fifteenth century and is completely covered with frescoes. Sadly, much of the magnificent forest surrounding it burned in 2008. So instead of a picnic in a glade, try a wine-tasting expedition at the vineyards round Embonas, which produce fine whites, reds and bubbly. But please do not visit the Valley of the Butterflies (a.k.a. tiger moths) - a surfeit of admirers has drastically reduced their numbers.

MANDRAKI HARBOUR

Rhodes

Emblematic bronze deer top the columns at the entrance to the old harbour where the legendary Colossus once stood. Or did he? No trace has ever been found and scientists question the feasibility of such a huge statue. Mandraki shelters yachts and tour boats today and looks on to the city designed by Mussolini's architects. Amongst the Fascist buildings in various monumental styles are a lovely mosque and Turkish cemetery, along with eateries featuring menus in Scandinavian languages. The big resort hotels form a chain along the town beach north of Mandraki.

THE OLD CITY

Rhodes

Unofficially, the Old City has replaced the Colossus as one of the world's wonders. A UNESCO World Heritage Site, it is the largest walled medieval town in Europe, its military architecture unsurpassed. It consists of three districts: the Collachium (castle) of the Knights, the restaurants and bazaar of the romantic Turkish quarter, and the residential, formerly Jewish, neighbourhood. The cobbled Avenue of the Knights flanked by seven Gothic inns of the Catholic Crusaders leads to the impeccably rebuilt Palace of the Grand Master. By Ottoman decree, the Greeks lived outside the walls.

LINDOS CASTLE

Rhodes

The Knights of St John were only in Rhodes from 1309 to 1522, but it was time enough to build enduring monuments. At Lindos they chose the ancient acropolis for their fortress. The jumble of ancient, medieval and Byzantine ruins plus a superb location has made the town a tourist mecca. Many jet-setters have villas here, protected from curious eyes by walled courtyards paved with mosaics of black and white pebbles. Some are on view in the fifteenth-century Panagia church, where rainbow-hued dresses and T-shirts flutter against the old white-walled lanes as the crowds press through.

TEMPLE OF ATHENA, LINDOS

Rhodes

Lindos was the most significant of the three city-states on Rhodes. Ialysos and Kameiros, on the west coast, are interesting though less impressive. This temple is invisible from below but the approach, though hard on the knees, could not be more dramatic. First you climb a steep path and a long stairway to the castle's main gate, then up several steps to a first-century BC terrace, where a monumental staircase climbs to a higher terrace. Only from the top step does the temple reveal itself entirely, seemingly floating between sea and sky.

THE APPROACH TO EMBORIOS

Halki

Halki's main road is called Tarpon Springs Boulevard. For centuries, farming being unrewarding among its rocks, its men fished for sponges, making enough money to construct grand houses on the waterfront. The collapse of the sponge industry sent them fleeing to America, especially Florida, in search of jobs. Practically deserted for years, Halki's resurgence started in 1983 when UNESCO proclaimed it 'an island of peace and friendship for international youth'. Since then, decaying stately homes are being renovated for tourists preferring quiet over crowds when they visit its castle, monasteries and beaches.

THE WATERFRONT

Kastellorizo

Just 2.5 km (1.5 miles) off the Turkish coast, Kastellorizo needs a special box in order to appear on maps of Greece. And although it is the smallest of the Dodecanese, the ancients named it Megiste ('Biggest') because the surrounding islets are even more minuscule. In 1910 it supported 9,000 inhabitants. Thirty years ago, most handsome mansions lay derelict, still scarred by World War II bombings. Today, islanders returning from overseas are making Kastellorizo an attractive, fun exception to the rule that every place we visit has seen better days.

YIALOS HARBOUR

Another island close to Turkey, Symi's main town resembles a less austere Hydra. Its houses rise steeply from the narrow port, dwindling in size as they ascend. Many streets are stepped, off limits to cars, but the neoclassical buildings have cheerful, pastel façades, not grey stone. Yialos, the port, merges seamlessly with Horio or Upper Symi, an older village with a castle at its core. At Pedhi, the landscape spills down the hill to the sea. Symi's beaches are only fair but it is undeniably photogenic and thus invaded by day-trippers from Rhodes.

BELL TOWER OF THE PANORMITIS MONASTERY

The eighteenth-century monastery of the Archangel Michael Panormitis is the island's main attraction for Greeks. A large complex on an enclosed bay, its guest hostel and restaurant are invariably filled with pilgrims come to pray to its miracle-working icon. Yachties can hear their excited voices well before they round the point to this excellent anchorage. The monastery has a folklore and ecclesiastical museum plus excellent frescoes. The bus back to Yialos passes through unexpected pine and cedar forests. Anatolian flavours – like hot pepper on fried shrimp– spice up local dishes.

TRADITIONAL FISHING DINGHIES

Tilos

In the 1970s, Tilos made news with the discovery of bones from a pygmy elephant that became extinct around 4600 BC. The huge head with gaping hole where the trunk had been gave rise to speculation that the beast could have inspired the Cyclops myth. Otherwise, Tilos is gratifyingly un-newsworthy. A harbour called Livadi ('Plain') and a capital called Megalo Horio ('Big Village') hint at what to expect: a delightful, tranquil village, hikes through pastures and orchards with views galore, unfrequented pebbly coves and fresh fish for lunch.

MANDRAKI

Nisyros

One of the first things you notice on approaching Nisyros are derelict baths under repair. Hot springs bubble on this volcanic island, and one spa stands near ruined Roman facilities. But before rushing to view the collapsed, still steaming crater, take a stroll round Mandraki's brightly painted houses. Then climb to the thirteenth-century castle, where the Panagia Spiliani church was created inside a cave, and follow signs to Paleokastro. You'll be surprised by a perfect classical fortress, with 19 courses of original masonry in stunning black stone.

MAIN SQUARE IN NIKEA

Nisyros

Two villages cling to the rim of the crater. Emborio is virtually abandoned, but Nikea is a gem, worth seeing just for its magnificent pebble-mosaic square. From here there are superb views of the volcano and Tilos in the distance. Keeping strictly to the path and wearing sturdy boots, you can walk into the mephitic moonscape, where sulphurous fumes cloud the atmosphere. The volcano last erupted in 1422, spewing out mud and gases but no lava. The island's rich volcanic soil produces superior fruit and vegetables.

OLD TOWN

Astypalea

Shaped like a butterfly with frayed wings, Astypalea belongs to the Dodecanese but looks and feels Cycladic. Requiring considerable effort to reach, it is so appealing you might not want to leave. Serene beaches are shaded by tamarisks, tavernas at Skala, the port, set tables in the sand, and the hora is a dream. There, sapphire and emerald woodwork alleviates the blinding whiteness, while rows of barrel-vaulted chapels, as many as five joined together, line the alleys. The castle at the top encloses an abandoned medieval village with two typically Cycladic domed churches.

GARDENS AND MINARET

Kos

Kos, with the second largest Castle of the Knights, also runs second after Rhodes as the most visited of the Dodecanese. Although Italians rebuilt the Old Town after an earthquake levelled it in 1933, some Anatolian features remain: impressive mosques, tall palms, even a sizeable Turkish community. The ancient city shares a site with the Knights' precinct, south of the castle proper. It features a Hellenistic agora with colonnade, temples and a fifth-century Christian basilica. Mosaic-floored Roman villas, three Roman baths and the acropolis make Kos a treat for archaeology buffs.

TEMPLE RUINS AT THE ASKLIPIEIO

Kos

The town's most celebrated landmarks are connected with its illustrious son, Hippocrates. Unsurprisingly, Kos draws many doctors' conventions. They pay homage to the immense plane tree under which Hippocrates is said to have taught and then visit the Asklipieio, dedicated to the god of medicine. After Hippocrates' death in 377 BC, the shrine (on three spacious terraces) functioned as a clinic and medical school until the sixth century AD. There physicians learned that illnesses have natural causes and can be treated with science not magic.

TYPICAL BEACH

The young people and Scandinavian sun-worshippers who descend on Kos each summer are probably more enticed by beaches than antiquities, not to mention bars and nightlife. Their favourite beaches lie to the southwest, along a huge crescent bay, with names like Magic, Sunny, Banana and, mysteriously, Camel. Inland, the leafy villages on the Swiss-looking slopes of Mt Dikeos for the most part lack the tourist trappings of the coast. The flat coast is one of the few places in Greece where riding a bicycle does not require Lance Armstrong's lungs.

AERIAL VIEW OF POTHIA

Kalymnos

The top sponge-fishing island, Kalymnos doesn't let you forget it. There are exhibits, divers' statues and two museums devoted to the industry in the port town of Pothia. It still gives a small fleet of divers a festive send-off every spring, although its processing plant now imports some of its sponges. Busy and noisy, its buildings are white, not coloured as on most of the Dodecanese. The Italians ordered painted façades. The Kalymniots refused. Stark mountains, green valleys, Kalymnos is a study in contrasts.

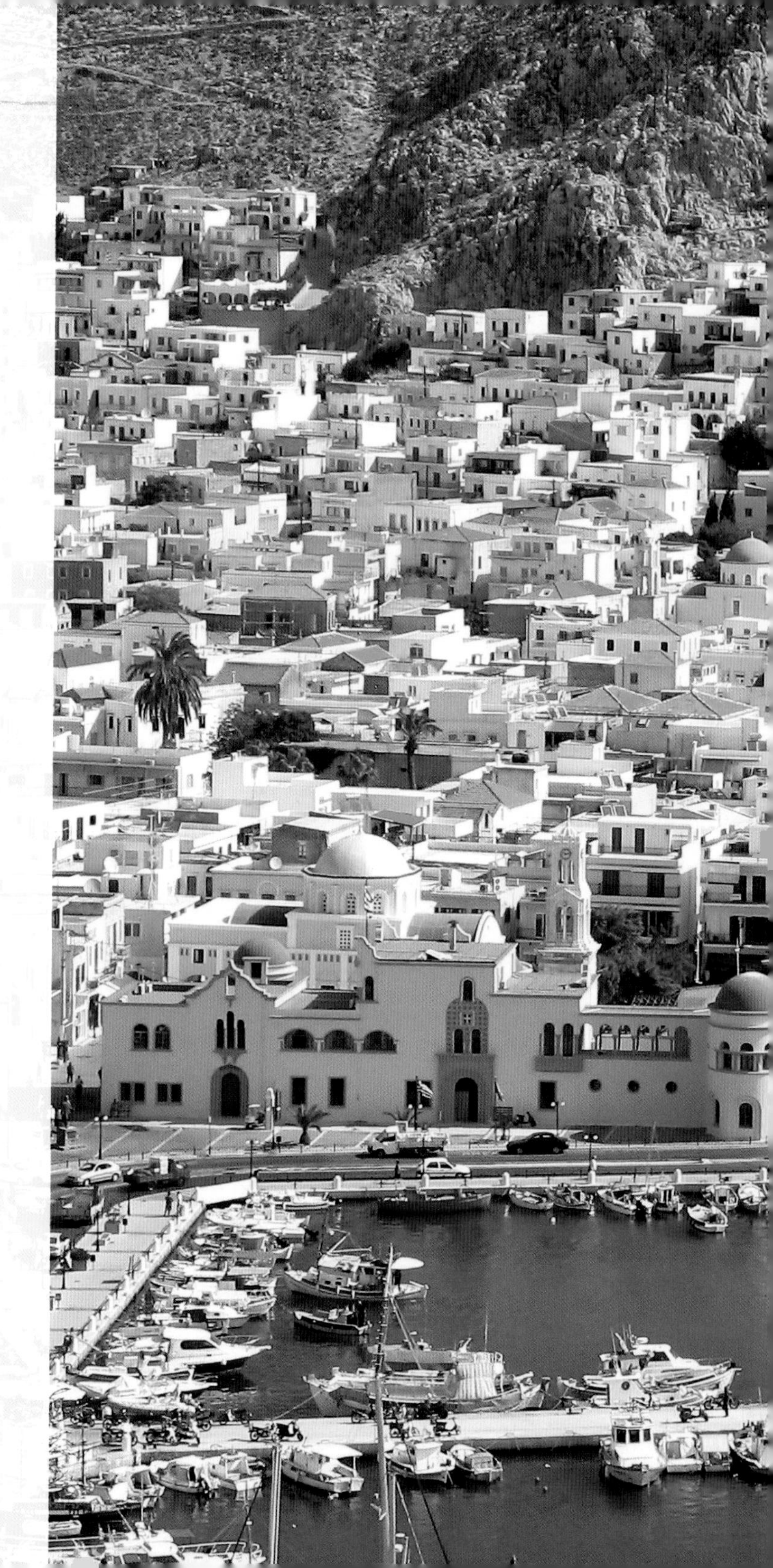

VIEW OF MASSOURI FROM TELENDOS

Kalymnos

One such contrast is Telendos. While separated from Kalymnos by a swimmable (700 m/2,000 ft) channel, it has none of the big hotels, mini-markets and fast-food restaurants that accompany the beach resorts at Massouri, Myrties and Armeos across the water. Instead, decades behind them, it offers glimpses into the recent and remote past, being littered with the remains of early Christian settlements that existed before a 14-day earthquake created the channel in 554. Literally dozens of basilicas from that era are scattered around Kalymnos, along with caves that have yielded Neolithic and Mycenaean finds.

WATERFRONT TAVERNA

Leros

Three peninsulas make up Leros and they are deeply indented with no fewer than seven bays. Thucydides mentions their strategic value and the Italians noticed it, too. The one at Lakki, the main port, is virtually enclosed, more lake than sea. The Italians made it their naval headquarters, endowing it with some of their best efforts at Fascist Art Deco. The broad, empty avenues and important buildings give Lakki a surreal atmosphere that speeds visitors away to more welcoming bays and villages, of which there are many, as yet undiscovered by tour operators.

141

SEA-GIRT CHAPEL

Leros

The charms of Leros are no secret to Greeks. One of its attractions is that there are no antiquities, no sites demanding inspection. Leros promises a guilt-free holiday, with taverna lunches at the water's edge, strolls to a picturesque chapel, swims from a shaded beach in clear, but chilly, water. At Platanos, the capital, 3 km (1.5 miles) north of Lakki, and Alinda, a modest resort below, neoclassical houses strike a gracious note, while the energetically inclined can climb to the crumbling fortress of Pandeli. Good Italian roads connect the other lovely bays.

THE PORT

There's not much to say about Lipsi. A speck in the Aegean between Leros and Patmos, it used to be known only to yachtspeople, who would spin yarns about perfect coves and friendly locals. Now excursion boats from both islands enable day-trippers to confirm those rumours. Lipsi has one village, whose main square is a wide circle with narrow streets radiating from it. Most tavernas and rooms are here or on the waterfront. Cars are few but those perfect coves are just 10 minutes' to an hour's walk away.

ΑΓ ΝΙΚΟΛΑΟΣ ΝΑ 55

AERIAL VIEW

Patmos

Many of the other Dodecanese may display Italian, Turkish or Frankish influences, but Patmos is pure Greek. Here the Orthodox Church has moulded its character, taking inspiration from a notable exile in AD 95. The Romans used to dispatch political prisoners to this stony place. One was St John the Divine who, taking refuge in a cave, took down the words of God. We know them as the Book of Revelation. In 1088 the Byzantine emperor Alexis Comnenos gave permission for a monastery to be founded in the saint's honour.

ENTRANCE COURT, MONASTERY OF ST JOHN

Patmos

Built as a massive fortress to deter pirates, the monastery crowns the hill above the port. The entrance court leads to dimly lit chapels with superb period frescoes, the refectory's long tables and the incomparable treasury. With a library containing thousands of volumes and documents, a sacristy crammed with icons, gold and silver, this collection has no equal in Greece outside Mt Athos. Though not open to the public, many impressive pieces are on display. A well-worn path from Skala passes the cave containing St John's stone pillow and table.

FISHING KAIKIS AT SKALA

Patmos

Back down at Skala, the port, Patmos loses some of the calm dignity conferred by the monastery and the immense, grey-shuttered mansions clustered round it. They date from the sixteenth and seventeenth centuries when the island's merchants and fleet brought great wealth. From here, fishing boats depart for the pleasant but not spectacular beaches hemming its ragged coast. One, Lambi, was renowned for its multicoloured stones. They proved irresistible souvenirs – none are left. Patmos has become a favourite with the discreetly rich and famous, who never mingle with the cruiseboat pilgrims.

AGIOS GEORGIOS

Agathonisi

With a population of 122 hardy souls, Agathonisi is even more minute than Lipsi. Yet it boasts a startling three villages: Agios Georgios and the descriptively named Megalo ('Big') and Mikro ('Little') Horio. But it does have a couple of *pensions*, eateries and quiet pebbly beaches. Together with Arki and Marathi, these are the ultimate getaway islets. Marathi has one family, but in summer its beach gets thronged with yachts and its taverna rings with laughter. Agathonisi and Arki fill to bursting on their saints' day festivals in July and August.

THE NORTH-EAST AEGEAN ISLANDS

These mostly large islands, with the exception of Thasos, hug the shores of Asia Minor. Their written history begins with colonization by post-Mycenaean 'refugees' from Attica and the Peloponnese known as Ionians, who also settled across the straits in what is now Turkey. Civilization shone here long before the Athenian miracle.

By the eighth century BC, Homer, claimed by Chios, was spinning his epics, by the 7th Sappho and Terpander were writing poetry and music in Lesvos, and Eupalinos and Pythagoras were establishing the basics of geometry in Samos. Further north, Limnos and Samothraki were the centres of worship of the Great Gods.

Mountainous yet fertile, strategically placed on trade routes between the Dardanelles, Egypt and the Holy Land, these islands have attracted the attention of many a conqueror: Persians, Athenians, Spartans, Romans, Byzantines, Arabs, Seljuk Turks, Venetians, Genoese, Ottomans and Mediterranean pirates of every stripe, all have been drawn here. Today, smatterings of this multicultural past resonate in often stunning scenery, while agriculture and shipping revenues have freed these islands from the need to court tourism. While foreigners are always welcome, only Samos and Thasos can be considered resort islands. Less wealthy, Ikaria and Samothraki appeal to off-beat tastes.

PYTHAGORIO

Samos

Samos reached its peak in the sixth century BC. Pythagorio, named after the renowned mathematician who was born here, occupies the site of the ancient city. Not far from its buzzing café/beach scene are some relics of that remarkable era. Most notable is the Eupalinos aqueduct/tunnel. More than 1,000 m (3,000 ft) long, it was calculated with such precision that its two halves, dug from opposite sides of the mountain, met in the middle. The feat has not been duplicated despite our 'advanced' techniques. The Persian invasions ended these glory years.

MARATHOKAMBOS

Samos

According to myth, Samos was Hera's birthplace. The Heraion, her sanctuary near Pythagorio, was one of the ancient world's Seven Wonders. Four times larger than the Parthenon, the temple's one remaining erect column cannot quite convey its former grandeur. Beyond it, a surprising collection of prehistoric bones and fossils at Mytilinii stretches our imaginations even further. The almost uninterrupted beaches on the wide bay of Marathokambos tempt us back to the present. Above them spreads the island's largest olive grove. Grapes grown on the appropriately named Mt Ambelos ('vine') produce its celebrated sweet wine.

KOKKARI BAY

Samos

Kokkari is a windsurfers' haunt on the north coast, not far from the capital Vathy. While most visitors hasten away from this busy port as soon as they arrive, the town deserves closer inspection. More village than city, the old upper district surveys the deep harbour from three sides, the public garden is a floral extravaganza, while the archaeological museum's statues and displays reveal a bit of the Heraion's wonder. Over a drink in a waterfront café, consider a day trip to Turkey. Ancient Ephesus lies just across the straits.

MIKRO SEITANI BEACH

Samos

You have to make an effort to get to this special beach and its larger sibling, Megalo Seitani, in northwest Samos. It's accessible only by footpath or fishing boat from Potami, where the road ends. There, twin white crescents and a waterfall might seduce you into staying. But try to resist the lure of the coast. The island's wooded mountain scenery should not be missed. Nightingales trill near a stream at Manolates, fruit drops from the trees at Vourliotes. Come at twilight after the foreigners depart and you'll glimpse the real Samos.

AGIOS KIRIKOS

Long and steep, Ikaria is as tough and rugged as Samos is lush and gentle. It falls administratively under the richer island and complains of getting less than it deserves. Over many centuries, it served as a place of exile, yet its people could not be more welcoming. They often lead double lives – as housepainters in New Jersey, taxi drivers back home. Ikarus plunged into the rough sea here; the Romans loved its healing waters. Around Agios Kirikos are a dozen hot springs, good for whatever ails you.

MAGICAL BEACH

Ikaria

Spa towns do not appeal to all, so most visitors hurry off to the more immediate charms of the north coast. Evdilos, 60 km (40 miles) of twisting, breathtaking mountain road away, is the largest port on this side. To the west of it magnificent beaches unfold. *Pensions* and tavernas have grown up around Kambos and Armenistis, but in a low-key manner that suits the young and frugal. Lacking antiquities and monuments, Ikaria is a night owl's paradise. In the mountain hamlets at Rahes, shops open in late afternoon and don't shut until 3 a.m.

THE MASTIC VILLAGE OF PYRGI

Chios

Chios is renowned for two things – its medieval mastic villages and the horrific massacre of 1822, portrayed in Delacroix's famous painting. After 1204, Chios fell under the sway of the Genoese. They built the castle on the port and then constructed 20 walled villages in the south, the only place in the world where the mastic tree exudes tears of precious, chewy resin. Used in gum and medicines since Hippocrates' day, it made both islanders and overlords prosperous. The 'xysta' decorations covering Pyrgi's walls are unique to Chios.

DESERTED VILLAGE OF ANAVATOS

Chios

The Sultan ousted the Genoese in 1566 but ruled with a gentle hand until 1822. When the Chiots, inspired by events on the mainland, rebelled against the Turks, that hand turned brutal. Estimates at the total slain range from 25,000 to 60,000 men, women and children. Anavatos has been abandoned since then, but the Sultan spared the mastic villages. You would find Chios far from gloomy, however, were you to visit the port's museums and markets, the amber-walled estates at Kambos, Nea Moni's eleventh-century mosaics and 101 other sights.

MYTILINI TOWN

Lesvos

Mytilini is not geared to tourism, which makes it that much more fun to explore. Brightly painted Ottoman-style mansions fronted by Ionic columns, a spacious green park, a warren of market streets and old-fashioned cafés meet in an agreeable jumble behind the waterfront, overlooked by the great, grey hulk of a Byzantine-Genoese castle. Outside the town, at Varia, two museums attract admirers of Picasso, Matisse and Theophilos, a beloved Greek naif painter. Plomari, down the coast, is the ouzo centre of Greece. The factories there offer tasting tours.

NM.347

MOLYVOS

Lesvos

To get to Molyvos (a.k.a. Mithymna) across the island you must pass Kalloni, at the back of its largest lagoon. Flamingos reside here, but so do the anchovies and sardines which gourmets relish, salted, with their ouzo. Picturesque Molyvos with its aristocratic houses snuggled under the hilltop castle has drawn tourists since it was a New Age magnet in the Seventies. Taverna tables crowd the waterfront, but its vine-shaded cobbled streets are car-free. Fishing boats deposit bathers at the fine beaches to the north. At Eftalou, hot springs bubble in the sea.

CHRISTINA.

AGIASSOS

Lesvos

Cool Agiassos is the centre of agricultural Lesvos, surrounded by a proverbial ocean of olive groves and forest. Pilgrims and pickups ring the cloistered main church, which gives on to the food 'souk', a row of stalls brimming with local specialities – cheeses, olives, sausages, honey – as well as farmers' cafés and potters' workshops. In contrast to the thickly wooded south, the western peninsula is wild and bare. Its most famous landmark is a petrified forest, best glimpsed at Sigri's excellent museum. Nearby, the Skala Eressou beach resort is famed for being Sappho's birthplace.

VIEW FROM CASTLE, MYRINA

Limnos

From the fort above the island's chief town of Myrina, you have a hawk's view of its two pretty waterfronts. Turkish Beach, shown here, is where the ships dock and fish tavernas cluster. Quieter Greek Beach, to the north, has handsome neoclassical homes, a museum and restaurants. Virtually ignored by tourists, domestic or foreign, owing to misguided perceptions that it is all wheat fields and military bases, Limnos repays visitors with fantastic food and wine, sand dunes, antiquities, cordial locals and, at Moudros's big but empty harbour, poignant reminders of its closeness to Gallipoli.

HORA

Samothraki

Samothraki holds nothing but surprises. The side facing Limnos could be a continuation of that island's tawny rolling hills, but go north and you'll find a riot of greenery, watered by as many as 20 rivers, more than there are roads or hamlets. Most outsiders stay at Therma, a trendy spa drenched in flowers. Half-deserted Hora has few amenities besides its spectacularly positioned cafés/tavernas. They offer good views of the formidable peaks of Mt Fengari, where Poseidon sat to watch the Trojan War.

SANCTUARY OF THE GREAT GODS

Samothraki

This may be one of the most satisfying ancient sites in Greece. Just enough survives to be evocative and the setting could not be lovelier. The Great Gods, or Kabeiroi, predated the denizens of Olympus. Their names are strange to us and were never uttered, but the Great Mother, Axieros, came to be identified with Demeter. The Hellenistic and Roman remains on the same site include a huge circular building, a palace and a theatre. The Parian-marble 'Winged Victory' decorated a fountain until it was stolen by a Frenchman in 1863.

FONIAS GORGE AND WATERFALL

Samothraki

No one goes to Samothraki for the swimming, unless they like rock pools and tumbling rivers. The antithesis of most Aegean islands, this one has only a couple of decent beaches. A more exciting prospect is to follow the trail along the Fonias ('killer') river among gargantuan, twisted plane trees up to these delicious falls. You won't be alone though. Flushed hikers loll on the smooth rocks beneath them like basking seals. Spring floods can cause the river to sweep everything in its path – goats, people, cars – out to sea.

PANAGIA VILLAGE
Thasos

If Samothraki suffers because it's so remote, Thasos's problems arise from being too accessible. Less than an hour from the Macedonian coast at Kavala, 30 minutes from Keramoti, this near-perfect and greenest of islands has been discovered by cheap tours from Athens, the UK and the Balkans. Panagia and its twin, Potamia, with their northern architecture and slate roofs, are close to Limenas (a.k.a. Thasos), the main town. There, roses and huge trees mask the agora and ancient theatre. Visitors buy honey in Panagia and then head for Golden Beach below.

LIMENARIA

Thasos

Like Samothraki, Thasos is round, but there the comparison stops. White beaches encircle it and pines and leafy trees cast shadows on translucent waters. The west coast is mild. After Limenaria, the halfway point, the scenery turns rugged and kiosks tempt drivers to stop to gaze at views of a convent clamped to a precipice or a tantalizing cove. Limenaria has another sort of drama. Straddling the hill between beach and port are three massive brick chimneys and palatial offices, romantic relics from early-twentieth-century mining operations.

A HOUSE IN THEOLOGOS

Thasos

Theologos was the island's capital in medieval and Ottoman times. Mass tourism has not compromised its traditional atmosphere and hospitality. Instead, it's the locals' hankering to be modern that is prompting them to replace centuries-old slate roofs with red tiles. Thankfully, the churches' wonderfully pointed, dunce-hat roofs have not been sacrificed. Down on the coast at Alyki is a splendid sight. At the tip of a wooded promontory lies a partly submerged ancient marble quarry, polished by wind and waves into surreal shapes, a single column drum presiding over the scene.

THE NORTHERN SPORADES & EVIA

So called because they are sporadically scattered in the Aegean (as opposed to the ring-like distribution of the Cyclades), the far-flung Northern Sporades start off the coast of the Mt Pelion peninsula.

Three of the main islands – Skiathos, Skopelos, Alonnisos – look like siblings, thickly wooded extensions of the parent peninsula to which they were once attached. Skyros, related neither geographically nor culturally, sits like an orphan off central Evia. In a nutshell, Skiathos is a party island, Skopelos aims at refinement, Alonnisos attracts nature lovers and Skyros holds on to its customs. Beyond them the National Marine Park, another 30 islets and 2,200 sq km (1,300 sq miles) of sea, offers protection to the endangered monk seal and other species.

Evia parallels the mainland so closely that some maintain it lacks island atmosphere. Nevertheless it is Greece's second largest, after Crete. Though relatively unknown abroad, it once supported three flourishing city-states – Halkis, Eretria and Karystos – and both the Venetians and Ottomans prized its resources. From the north's classy spas and thick forests to the south's trekking terrain and unspoilt mountain villages, from the west's calm lagoon-like coast to the east's wilder beaches, from Halkis's sublime seafood to Eretria's antiquities, Evia makes one promise: seek and you will find.

VIEW FROM HORA

Alonnisos

If it had been taken 30 years ago, this photo's houses would have had no roofs. Cracked walls and rubble were all that remained of Hora after a powerful earthquake in 1965. Overnight it turned into a waterless ghost town, as inhabitants were forced to move into hurriedly built cement blocks on the coast. Today Hora has been coaxed back to life, mostly by Germans, and in summer its galleries, tavernas and bars are thronged with international locals and tourists come to gasp at the view of all the Sporades from its castle ramparts.

FLYING DOLPHINS
CHRISTINA
AKTH
SIXT
AVIS
ROOMS TO LET

PATITIRI

Alonnisos

Patitiri means wine press, but in the Fifties ill-fated Alonnisos lost its vines to the pest phylloxera. Its buildings are ordinary at best, but the white cliffs studded with pines and the bottle-green water around the little harbour and nearby coves improve its aspect. Ubiquitous flowers and the islanders' real charm soon have you relishing the funky ambience. Besides, if you're not cruising by on a yacht, you've come here to chill out on pine-scented beaches, dine on just-caught fish and perhaps see the seal pups at the refuge at Steni Vala.

SKOPELOS TOWN AND PORT

Skopelos

Placed between Alonnisos and Skiathos, this island suffered none of their twentieth-century disasters (Nazis punished Skiathos's resistance by burning its main town). And the absence of an airport has spared it from the worst excesses of mass tourism. Skopelos watched the changes next door and vowed not to repeat them. As a result, Hora, once you leave the port, retains its village atmosphere. Tall Pelionesque houses shade almost vertical lanes, interspersed with dazzling old churches, until you reach the castle built on ancient foundations and a most welcome ouzeri for gazing at the view.

GLOSSA

Skopelos

Glossa stands above the island's other port, Loutraki, opposite Skiathos. It's a quieter, even less commercial version of Hora. The islanders' respect for tradition shows in small things like hand-painted signs above shops, artisans' boutiques (instead of furriers, jewellers and mass-produced knick-knacks), imaginative menus in sophisticated tavernas and well-marked hiking trails. From Hora, four imposing monasteries hidden amongst thick greenery await inspection on the other side of the harbour; from Glossa, a path leads through plum orchards to Ai Yanni Kastri, the wave-bashed chapel in *Mamma Mia*.

KASTANI BEACH

Skopelos

Almost all the island's beaches face south, like this one near Glossa. Though they are beautiful, they can be counted on two hands and tend to have more pebbles than sand. This relative deficiency has saved Skopelos from turning into one big bather's paradise like Skiathos, whose smaller coastline is scalloped with 60 to 70 beaches of varying size. Still, the choice is ample enough, ranging from white sand and watersports at Milia to coves for nudists at Velanio. Besides, there are more than 360 churches and 40 monasteries to photograph.

HOUSES IN HARBOUR

Skiathos

Skiathos's only town covers two low hills overlooking two waterfronts, one busy and featureless, the other devoted to cafés, ouzeris and people-watching. Pubs, *pensions*, glamorous pool bars, fast-food joints, tavernas, discos and shops selling everything from naughty postcards to kilims and antiques squeeze into every available space on the back streets. Lacking traditional architecture, this eclectic collection radiates fun and good humour, while flowering vines smother most alien constructions. Hotels, rooms and eateries crowd behind the beaches all the way to Koukounaries, once dubbed Greece's best beach, now far too developed.

EVANGELISTRIA MONASTERY

Skiathos

This graceful eighteenth-century church belongs to a monastery above the town. From here it's a lovely, non-taxing walk to Kastro, the walled, cliff-side settlement where the population eked out an uncomfortable existence from 1540 to 1829. It had 30 churches – three since restored – and 300 houses. Tour boats also dock here on excursions round the island, passing a row of grottoes hollowed from the rocky shore. A high point of their tour is Lalaria, the island's second most renowned beach, preceded by a towering arched rock. Tourists find its smooth, round stones make good souvenirs.

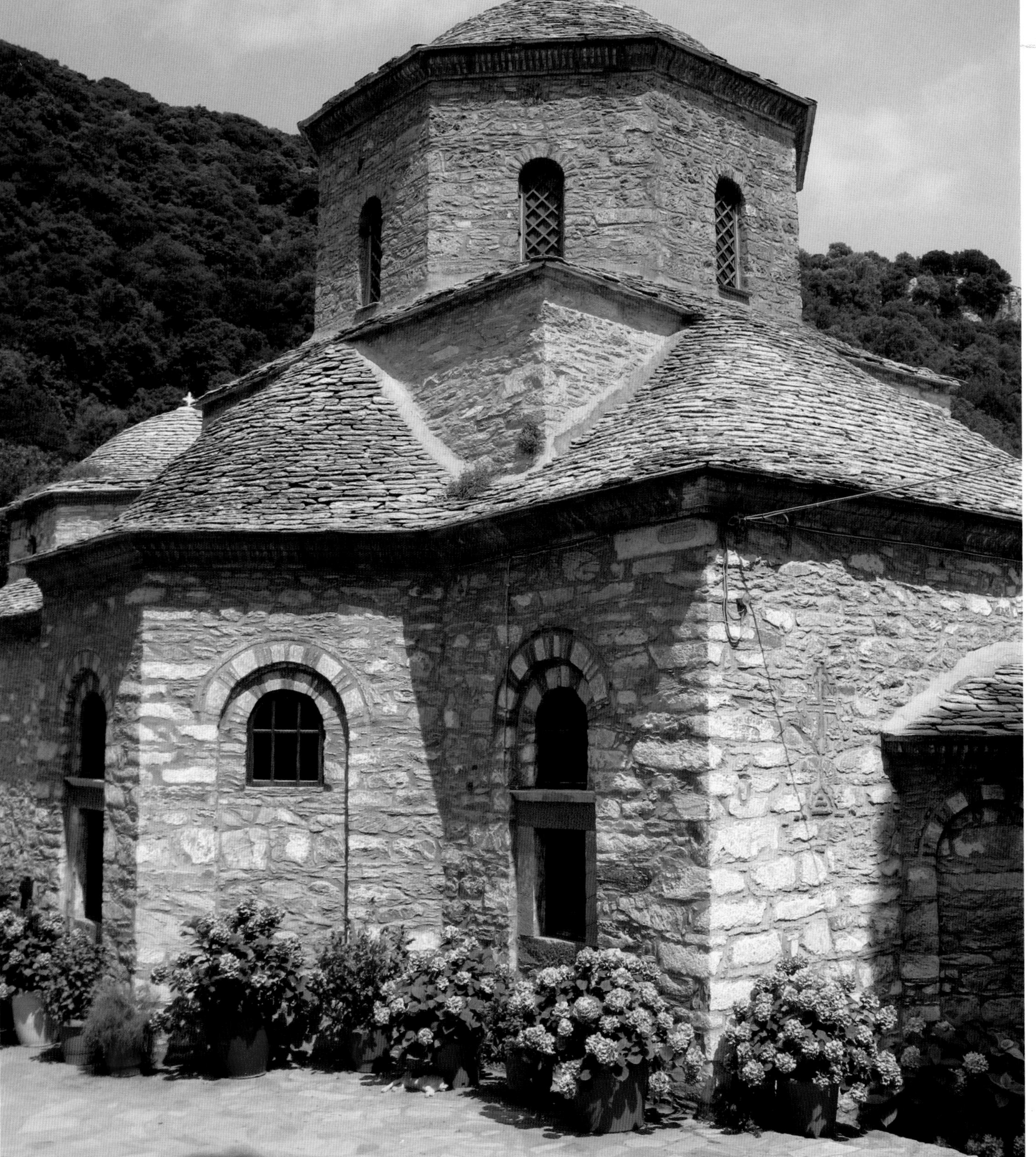

HORIO

Skyros

Horio, the capital, has been inhabited since time immemorial. The Mycenean-era kastro-acropolis at its peak is where Achilles hid from Odysseus before the Trojan War. Isolated even today, Skyrian customs have resisted 'kitschification'. Here charmingly painted ceramics, small carved chairs and delicate embroidery are made with the same care as in the past. Every house is a museum. Below Horio, beach tavernas at Magazia and Molos serve lobster. Elsewhere, Rupert Brooke's tomb – a corner that is 'forever England' – and a New Age centre attract tourists of a different sort.

MT OHI WITH SNOW

Evia

This peak dominates southern Evia. It could be called the Magic Mountain. Hera and Zeus were married here, and a mysterious structure of enormous slabs near the summit may be an eighth-century BC temple to the goddess. The locals call it the Dragon House. Lower down, unfinished columns from a Roman quarry litter the slope. Walks abound. The wooded Dimosaris gorge runs down to the Cavo d'Oro straits separating Evia and Andros. Paths link Turkish fountains, old stone bridges, Byzantine churches, a Roman aqueduct, a Venetian fort and several well-watered villages.

CHURCH AT KARYSTOS

Evia

Below Ohi, Karystos spreads out behind a large, sheltered bay with an endless beach. Despite this privileged location, the town is no resort. Hardware stores selling donkey beads and cheese-baskets, old-fashioned confectioners, and ouzeris with dangling octopus tentacles line a perfect grid of streets, not the usual island maze. King Otto loved it and had his Bavarian architects lay it out and build some neoclassical gems. It was called Othonoupolis until his deposition in 1862. With its pleasant hotels, genuine tavernas and small museum, Karystos makes a perfect first stop in Evia.

INDEX